CW01464316

Microprocessor Engineering

B. Holdsworth

Butterworths

London Boston Durban Singapore Sydney Toronto Wellington

First published 1987

© **Butterworth & Co. (Publishers) Ltd, 1987**

British Library Cataloguing in Publication Data

Holdsworth, Brian
   Microprocessor engineering.
   1. Microprocessors
   I. Title
   621.391'6   TK7895.M5
   ISBN 0-408-01361-3

Library of Congress Cataloging in Publication Data

Holdsworth, B. (Brian)
   Microprocessor engineering.

   Bibliography: p.
   Includes index.
   1. Microprocessors.   2. INTEL 8085A (Microprocessor)
   I. Title.
   TK7895.M5H64   1986   004.16   86-17123
   ISBN 0-408-01361-3

Photoset by Butterworths Litho Preparation Department
Printed and bound in Great Britain by Anchor Brendon Ltd,
Tiptree, Essex

Preface

This volume has been developed from the material given in the undergraduate course in the Electronics Department at Chelsea College, and to practising engineers and scientists who have attended one-week courses which have been held at Chelsea from time to time over the last few years. Essentially, the book is an undergraduate text which should prove useful to students in Electronic/Electrical Engineering and Computer Science Departments. It may also prove useful to other types of undergraduate engineers and physicists who find that they are required to use microprocessors in project work at the third year level. In recent years microprocessors have also found their way into the schools in increasing numbers, and consequently this book may be found useful at the sixth form level or possibly earlier. Alternatively, practising engineers with no previous digital experience and dedicated electronic hobbyists may find something of practical value to them in the text.

With the advent of integrated circuit technology, computing facilities have become available to people at all levels of society, and, in particular, to people who have a professional or spare time interest. In recent years we have seen the development of the home computer market, and although that market may now have peaked, it is clear that in future there will be a far greater number of people than heretofore who will be able to take an interest in computing. By virtue of the powerful influence exerted by the media, there has been a huge flood of people who have entered this market. However, the engineer and the scientist need to delve somewhat more deeply than the ordinary man or woman in the street, and this volume has been written to provide a somewhat deeper insight of the structure and operating techniques of a small computing machine as well as a basic understanding of assembly language programming. Understanding these various aspects of computing does not require any great mathematical ability, nor does it require a deep knowledge of the fundamental principles of electronic engineering. All that is required is a good deal of common sense and an ability to deal with problems in a logical and orderly fashion.

There are two main aspects of microprocessors of significant interest to the practising engineer, namely hardware and software. Hardware requires a sound knowledge of the digital techniques described in *Digital Logic Design* (Holdsworth – Butterworths, 1982); for example, a good understanding of logical and arithmetic processes, a knowledge of the facilities provided by modern MSI and LSI chips, and the ability to interconnect the various parts of a microcomputer system including its interfacing with the analogue world. Additionally, he/she will require a sound working knowledge of the instruction set of the selected machine and will have to develop the ability to write programs in a structured and economic way.

The book is divided into ten chapters. The first two chapters cover basic arithmetic and logic processes. Chapter 1 outlines the various number systems commonly encountered, namely decimal, binary and hexadecimal, and then develops the techniques used for executing various arithmetic processes such as addition, subtraction, multiplication and division in a machine. Chapter 2 deals with the various gates and flip-flops used in microprocessor systems and goes on to introduce a variety of MSI chips such as registers, multiplexers, encoders, etc. which are commonly available and may be required as support chips in microcomputing systems. The third chapter deals with memory devices, both ROM and RWM, and discusses in an elementary fashion the circuit techniques employed in the two basic technologies. However, more emphasis is laid on memory organization, and the addressing techniques used for accessing memory. The final chapter in this group, Chapter 4, deals first with microprocessor architecture in general, but, in particular, a complete description of the structure of the Intel 8085A is given, and some space is devoted to its operation.

The next group of three chapters is concerned with the software aspects of machine operation. Addressing modes employed with microprocessors are dealt with in the early part of Chapter 5, while the latter half of this chapter deals with the data transfer, arithmetic and logic instructions available in the 8085A instruction set. In Chapter 6, flowcharts are introduced as a programming tool, and this leads to the concept of branching and the use of the jump instruction. The stack and the instructions associated with its control leads to the topic of subroutines and the use of the call and return instructions. Chapter 7 deals with the broader aspects of assembly language programming and briefly introduces some of the more sophisticated software aids.

Chapters 8, 9 and 10 discuss the relationship of the microprocessor with the outside world. For example, Chapter 8 deals with both parallel and serial I/O operations and a variety of other I/O topics such as polling and vectoring. In Chapter 9, interrupts are discussed in general terms and later in the chapter with particular reference to the 8085A. Finally, Chapter 10 deals with the transformation of both analogue and digital signals in some detail.

The author wishes to thank a number of people for the assistance they have given to the production of this book. First, I would like to thank two of my former students, Dr John Porter and Dr Paul Turney, both of whom have read the manuscript and made helpful suggestions. Next, I would like to thank all those students who have passed through my hands in recent years and who, by virtue of their enquiring minds, have helped to clarify my own thoughts. Finally, I would like to thank my wife, June, and my youngest daughter, Penny, for typing the manuscript and also for helping me in a multitude of ways. Without their tolerance and understanding the book would never have been written.

B.H.

Contents

1 Arithmetic

1.1 Introduction

A microprocessor-based computing system has to have a numerical computation capability, and consequently it must be able to implement the four basic arithmetic processes of addition, subtraction, multiplication and division. Human beings perform arithmetic operations using the decimal number system, but, by comparison, a microprocessor is inherently a binary machine and its numerical calculations are executed using a binary number system. Additionally, the hexadecimal system is used in program listings, so the engineer who proposes to use this kind of machine must be familiar with a variety of different number systems.

It is also true that arithmetic processes executed by a digital machine are not by any means identical to the pencil and paper methods normally employed by man. For example, the process of subtraction in this kind of machine is carried out as an addition, and this involves the use of complement arithmetic.

Again, a frequent requirement is that the output of a microprocessor should be a decimal display, for obvious reasons. Since the machine works in binary, a way has to be found to represent decimal numbers in terms of binary digits and this involves the use of binary coded decimal. Methods then have to be devised so that arithmetic operations in the machine can be performed using binary coded decimal, and techniques for converting from pure binary to binary coded decimal and vice versa are also required.

The purpose of this chapter is to describe the various number systems in common usage in microprocessor systems and to develop methods for implementing the various arithmetic operations that have to be performed.

1.2 Number systems

The number system most familiar to man is the decimal system. A decimal number such as $(473.85)_{10}$ may be expressed in the following form:

$$(N)_{10} = 4 \times 10^2 + 7 \times 10^1 + 3 \times 10^0 + 8 \times 10^{-1} + 5 \times 10^{-2}$$

An examination of this representation shows that the number consists of a series of decimal digits multiplied by the number $(10)_{10}$ raised to some power that depends upon the position of the decimal digit in the number.

The number $(10)_{10}$ is called the base or radix of the system and it is equal to the number of digits in the system. For example, in the decimal system there are ten digits, 0 to 9 inclusive. The base is the distinguishing feature between two number systems. The binary system for example has a base of 2 and consequently it only has two digits, 0 and 1 respectively.

The decimal magnitude $(N)_{10}$ of a number in any system can be expressed by the equation:

$$(N)_{10} = a_{n-1}b^{n-1} + a_{n-2}b^{n-2} + \ldots a_1b^1 + a_0b^0 + a_{-1}b^{-1} + a_{-2}b^{-2} + \ldots + a_{-m}b^{-m}$$

where n is the number of integral digits and m the number of fractional digits. The base of the system is b, and a is a positional coefficient. With the aid of this equation, the binary number $(1011.11)_2$ can be evaluated as shown in the following example.

$$
\begin{aligned}
(N)_{10} &= 1 \times 2^3 + 0 \times 2^2 + 1 \times 2^1 + 1 \times 2^0 + 1 \times 2^{-1} + 1 \times 2^{-2} \\
&= 8.0 + 0.0 + 2.0 + 1.0 + 0.5 + 0.25 \\
&= (11.75)_{10}
\end{aligned}
$$

The binary number system is important to the digital engineer because all digital machines operate with the binary system. In the field of electronic engineering there are many switching devices that have two clearly defined states, for example the bipolar transistor (BPT) and the field effect transistor (FET). One of these two states is used to represent the binary digit 1 and the other the binary digit 0. A BPT can typically operate with its collector-emitter voltage $v_{CE} \approx 5.0$ volts when it is non-conducting and it is common practice to define this condition as representing the binary digit 1.

Alternatively, when the transistor is in the saturation condition it can operate with $v_{CE} \approx 0$ and this condition can be used to represent the binary digit 0. It is relatively easy to switch the transistor between these two conditions when it is operating in a switching mode.

Two other number systems of some importance to the digital engineer are the octal, or base 8, system, and the hexadecimal, or base 16, system. The octal system has eight digits, 0 to 7 inclusive, and a typical octal number is $(27.2)_8$, whose decimal value is given by

$$(N)_{10} = 2 \times 8^1 + 7 \times 8^0 + 2 \times 8^{-1}$$
$$= 16.0 + 7.0 + 0.25$$
$$= (23.25)_{10}$$

The hexadecimal system has 16 digits and since there are only ten digits available it is necessary to invent some additional ones. The additional digits are by convention represented by the first six letters of the alphabet, A to F inclusive, and the decimal values of these additional digits are:

$$(A)_{16} = (10)_{10} \quad (B)_{16} = (11)_{10} \quad (C)_{16} = (12)_{10}$$
$$(D)_{16} = (13)_{10} \quad (E)_{16} = (14)_{10} \quad (F)_{16} = (15)_{10}$$

A typical hexadecimal number is $(A2.C)_{16}$ and the decimal value of this number is given by

$$(N)_{10} = A \times 16^1 + 2 \times 16^0 + C \times 16^{-1}$$
$$= 160.0 + 2.0 + 0.75$$
$$= (162.75)_{10}$$

1.3 Conversion between number systems

A number in any base can be divided into two parts: (a) the integral part to the left of the radix point, and (b) the fractional part to the right of the radix point. The process of conversion to another base is different for the two parts of the number.

The decimal value of the integral part N_I of a base b number is given by

$$(N_I)_{10} = a_{n-1}b^{n-1} + a_{n-2}b^{n-2} + \ldots a_2 b^2 + a_1 b^1 + a_0 b^0$$

Dividing both sides of this equation by the base gives

$$\frac{(N_I)_{10}}{b} = a_{n-1}b^{n-2} + a_{n-2}b^{n-3} + \ldots a_2 b^1 + a_1 b^0 + \frac{a_0}{b}$$

The result of dividing by the base is to leave the least significant digit of the number, a_0, as the remainder. Subsequent repeated divisions will produce remainders of $a_1, a_2, \ldots a_{n-1}$. As an example of the process of repeated division by the required base the decimal number $(100)_{10}$ is converted to its binary, octal and hexadecimal equivalents below:

```
2 | 100   0       8 | 100   4 ↑      16 | 100   4 ↑
2 |  50   0       8 |  12   4 |      16 |   6   6 |
2 |  25   1 ↑     8 |   1   1 |           0
2 |  12   0 |         0
2 |   6   0 |
2 |   3   1
2 |   1   1
      0
```

$$(100)_{10} = (1100100)_2 \qquad = (144)_8 \qquad\qquad = (64)_{16}$$

The decimal value of the fractional part N_F of a base b number is given by

$$(N_F)_{10} = a_{-1}b^{-1} + a_{-2}b^{-2} + \ldots + a_{-m}b^{-m}$$

and, if both sides are multiplied by the base, the equation becomes

$$b\,(N_F)_{10} = a_{-1} + a_{-2}b^{-1} + \ldots + a_{-m}b^{-(m-1)}$$

Clearly, the first multiplication reveals the coefficient a_{-1}. Subsequent multiplication will reveal the coefficients $a_{-2}, a_{-3}, \ldots a_{-m}$. As an example of this process the decimal number $(0.265)_{10}$ is converted below to its corresponding binary, octal and hexadecimal forms:

```
   .265 × 2          .265 × 8           .265 × 16
  0.530 × 2         2.120 × 8          4.240 × 16
  1.060 × 2         0.960 × 8          3.840 × 16
  0.120 × 2         7.680 × 8         13.440 × 16
↓ 0.240 × 2       ↓ 5.440 × 8        ↓ 7.040 × 16
  0.480             3.520              0.640
```

$$(0.265)_{10} = (0.01000)_2 \quad = (0.20753)_8 \quad = (0.43D70)_{16}$$

The number $(0.265)_{10}$ is expressed to five binary, octal and hexadecimal places respectively.

Besides these conversions from decimal to the three number systems, binary, octal and hexadecimal, it is relatively easy to convert from both octal and hexadecimal to the binary system.

The octal digits from 0 to 7 inclusive can each be represented by three binary digits as shown in Figure 1.1. To find the octal representation of a string of binary digits, the string is divided into

Octal	Binary
0	000
1	001
2	010
3	011
4	100
5	101
6	110
7	111

Figure 1.1 Octal/binary conversion table

groups of three, beginning from the right-hand end. The octal equivalent for each group of three digits is then written down with the aid of the conversion table as shown below:

$$(|110 \mid 001 \mid 011 \mid 100|)_2$$
$$= \quad (6 \quad\ 1 \quad\ 3 \quad\ 4)_8$$

Similarly, octal numbers can be converted to binary by replacing each octal digit with the corresponding three binary digits from the conversion table. For example:

$$(4 \quad 3 \quad 2 \quad 7)_8$$
$$= (100 \quad 011 \quad 010 \quad 111)_2$$

In the same way each of the 16 hexadecimal digits can be represented by four binary digits, as shown in the table in Figure 1.2. To convert a string of binary digits into their hexadecimal

HD	Binary	HD	Binary
0	0000	8	1000
1	0001	9	1001
2	0010	A	1010
3	0011	B	1011
4	0100	C	1100
5	0101	D	1101
6	0110	E	1110
7	0111	F	1111

Figure 1.2 Hexadecimal/binary conversion table

representation the string is divided into groups of four, beginning from the right-hand end, and each group of four binary digits is replaced by its hexadecimal equivalent from the conversion table, as shown below:

$$(| 1011 \mid 1010 \mid 0011 \mid 0010|)_2$$
$$= \quad (B \quad\ A \quad\ 3 \quad\ 2)_{16}$$

For the reverse conversion, each hexadecimal digit can be replaced by the appropriate four binary digits from the conversion table of Figure 1.2. For example:

$$(4 \quad F \quad C \quad 2)_{16}$$
$$\downarrow \quad \downarrow \quad \downarrow \quad \downarrow$$
$$= (0100 \quad 1111 \quad 1100 \quad 0010)_2$$

1.4 Representation of numerical data in a machine

Numerical data in a digital machine is represented in binary form by a string of binary digits. This string is divided into groups and each group is referred to as a word. Word length is one of the most important parameters of a digital machine and is closely related to its computing power. Common word lengths in the microcomputer world are 4, 8, 16 and 32 bits. In recent years the 8-bit machine has been widely used in practice. A word length of 8 bits is usually referred to as a byte. More recently, machines with a word length of 16 bits have been introduced.

Computer words are stored in registers. A register can be implemented by an array of flip-flops where each flip-flop can be regarded as a single-bit memory. Functionally, a digital machine can be regarded as a device consisting largely of registers, and data is moved from register to register during the course of arithmetic and logic operations. This is undoubtedly a simplistic view of a microcomputer and its operation, but nevertheless it gives a good basic picture of the machine.

The word length of a machine is associated with computing time. For example, a 16-bit machine can add two 16-bit numbers in a single operation; however, with an 8-bit machine each number would have to be split into two bytes. Addition of the two least significant bytes would take place first, followed by addition of the two most significant bytes, and the time taken to perform 16-bit addition on an 8-bit machine would by necessity be longer. This type of operation would be referred to as multi-byte operation.

1.5 Unsigned arithmetic

Addition

The rules for the addition of two single-bit numbers are defined by the table shown in Figure 1.3. The addition of two unsigned 8-bit numbers using the rules of Figure 1.3 is shown below:

Augend	Addend	Sum	Carry
0	0	0	0
0	1	1	0
1	0	1	0
1	1	0	1

Figure 1.3 Rules for the addition of two binary digits

	2^7	2^6	2^5	2^4	2^3	2^2	2^1	2^0	
Augend	0	0	0	1	0	1	1	0	22
Addend	0	0	1	0	1	1	0	1	45
Sum	0	1	0	0	0	0	1	1	67
Carries		1	1	1	1				

The weighting of the individual digits in the two 8-bit numbers is shown above the digits. As long as the sum of the two unsigned numbers is $\leqslant (255)_{10}$ it can be represented by a single byte.

In general, when two n-bit numbers are added together it is possible to obtain an $(n + 1)$ bit sum. This additional bit generated by the addition is called *arithmetic overflow*. When carrying out a paper and pencil calculation, the production of this extra bit creates no problem. However, in an 8-bit digital machine, the augend and addend are both stored in 8-bit registers prior to the addition, and after the addition has been performed the sum is returned to one of these two registers. In the event of a 9-bit sum being produced, an extra 1-bit register has to be provided to store the ninth digit. To overcome the above problem some 8-bit microprocessors such as the 8085A provide an instruction which, when executed, implements 16-bit addition.

An additional problem is also created by such an overflow. If the sum is to be stored in memory it will require two memory locations, since a single memory location can only store one byte. Hence, when single-byte addition is performed in a machine such as the Intel 8085A and a 9-bit answer is obtained, a carry flip-flop (1-bit register) is set and two memory locations have to be provided to store the sum.

Subtraction

The rules for subtraction are summarized in the table shown in Figure 1.4.

Minuend	Subtrahend	Difference	Borrow
0	0	0	0
0	1	1	1
1	0	1	0
1	1	0	0

Figure 1.4 Rules for the subtraction of two binary digits

An example of the subtraction of two 8-bit unsigned numbers is shown below:

	2^7	2^6	2^5	2^4	2^3	2^2	2^1	2^0	
Minuend	0	0	1	0	1	1	0	1	45
Subtrahend	0	0	0	1	1	1	1	0	30
Difference	0	0	0	0	1	1	1	1	15
Borrows				1	1	1	1		

If the subtrahend is greater than the minuend, an *arithmetic underflow* occurs which results in a borrow out at the most significant bit, as illustrated in the example shown below:

	2^7	2^6	2^5	2^4	2^3	2^2	2^1	2^0	
Minuend	0	0	1	1	1	1	0	0	60
Subtrahend	1	0	1	1	0	0	0	1	177
Difference	1	0	0	0	1	0	1	1	$\neq -117$
Borrows	1					1	1		

The difference in this case is a negative number, which is indicated by the borrow out occurring at the most significant place. The borrow out, or underflow, can be used to set a 1-bit register in a machine which, when examined, will reveal that the answer is negative. Unfortunately, when using unsigned arithmetic in a machine, there is no method of storing the difference as a negative quantity as the above example demonstrates, and for this reason unsigned arithmetic cannot be used to perform the subtraction process if the subtrahend is greater than the minuend.

Multiplication

The rules for binary multiplication are given in tabular form in Figure 1.5. An example of the multiplication of two 4-bit numbers is given below:

Multiplicand	1 0 1 1	11
Multiplier	1 1 0 1	13

```
          1 0 1 1  ⎫
          0 0 0 0  ⎪ Partial
        1 0 1 1    ⎬ products
      1 0 1 1      ⎭
```

| Product | 1 0 0 0 1 1 1 1 | 143 |

Multiplicand	Multiplier	Product
0	0	0
0	1	0
1	0	0
1	1	1

Figure 1.5 Rules for binary multiplication

It will be noticed from the above example that if two 4-bit unsigned numbers are multiplied together an 8-bit answer is obtained, and it is a general rule that if an m-bit unsigned number and an n-bit unsigned number are multiplied together, the resulting product will contain $(m + n)$ bits.

A set of rules for the process of multiplication can be stated as follows:

(1) If the least significant bit (LSB) of the multiplier is 1 write down the multiplicand and shift one place left.
(2) If the LSB of the multiplier is 0 write down a number of 0's equal to the number of bits in the multiplicand and shift one place left.
(3) For each bit of the multiplier repeat either (1) or (2).
(4) Add all the partial products to form the final product.

Such a set of rules is frequently called an *algorithm*. Algorithms can be implemented either in hardware or software. In a microprocessor-based system it would be normal practice to look for a software implementation of these rules unless time is a critical factor.

It is perhaps also worth noting that the multiplication process when executed by a microprocessor is performed in very much the same way as in the pencil and paper method. However, it differs in one important aspect; the partial products are accumulated as they are generated rather than all being added together at the end.

Division

The rules for the division of two single-bit numbers are summarized in the table shown in Figure 1.6. The division process can be regarded as one of repeated subtraction of the divisor X from the dividend Y. The number of times the divisor can be

Dividend	Divisor	Quotient	Remainder
0	0	Indeterminate	Indeterminate
0	1	0	1
1	0	∞	Indeterminate
1	1	1	0

Figure 1.6 Rules for binary division

subtracted is the quotient Q and the residue after the last subtraction is the remainder R, where $R < X$.

The division equation may be written

$$Y = QX + R \qquad R < X$$

where the quotient can be expressed as an n-bit word

$$Q = Q_{n-1}2^{n-1} + Q_{n-2}2^{n-2} + \ldots Q_1 2^1 + Q_0 2^0$$

and hence the division equation becomes, after substitution for Q,

$$Y = (Q_{n-1}2^{n-1} + Q_{n-2}2^{n-2} + \ldots Q_1 2^1 + Q_0 2^0)\, X + R$$

This equation defines the algorithm for division. Each of the terms such as $2^{n-1}X$, $2^{n-2}X \ldots 2^0 X$ is subtracted in turn from the dividend. After each subtraction a corresponding quotient bit and a partial remainder are obtained. For example, if $Y = 1101$ and $X = 0010$ and a 4-bit quotient is required ($n = 4$) then the first subtraction will be

$$Y - 2^3 X = 1101 - 10000$$

Clearly, subtraction cannot take place since $2^3 X > Y$, hence $Q_3 = 0$ and the first partial remainder $R_3 = Y$. The second subtraction is

$$Y - 2^2 X = 1101 - 1000 = 0101$$

In this case subtraction is possible, hence $Q_2 = 1$ and the partial remainder $R_2 = 0101$.

The third subtraction is

$$R_2 - 2^1 X = 0101 - 0100 = 0001$$

giving $Q_1 = 1$ and $R_1 = 0001$ and the final subtraction is

$$R_1 - 2^0 X = 0001 - 0010$$

Again, in this case subtraction is not possible, hence $Q_0 = 0$ and the partial remainder $R_0 = 0001$.

Since the division is to be taken no further, the quotient $Q = Q_1 Q_2 Q_3 Q_4$ and is given by $Q = 0110$ while the partial remainder $R = R_0$ and $R = 0001$.

In practice the machine implementation of the division process is slightly different from the method described above. It differs in that the dividend is divided by 2^{n-1} and is then compared with the divisor. If the divisor is less than the dividend divided by 2^{n-1} then subtraction takes place; if not, zero is subtracted from the dividend. In both cases a shift to the right takes place after subtraction. An example of the division process as performed in a digital machine is given below:

```
                        0  1  1  0    Quotient
Divisor   0  0  1  0 | 0  0  0  1  1  0  1 Dividend
                       0  0  0  0
                       ─────────
                          0  0  1  1
                          0  0  1  0
                          ─────────
                             0  0  1  0
                             0  0  1  0
                             ─────────
                             0  0  0  0  1 Remainder
```

1.6 Signed arithmetic

It is important that there should be a distinction made between positive and negative numbers in a digital machine. A sign digit is used to provide this distinction. A negative number is identified by a 1 that appears in the most significant position, hence

$-23 = 1, 0\ 0\ 1\ 0\ 1\ 1\ 1$

Conversely, a positive number is identified by a 0 that appears in the most significant bit position, so that

$+23 = 0, 0\ 0\ 1\ 0\ 1\ 1\ 1$

This type of representation is referred to as sign and magnitude representation. Because machine computation is difficult when numbers are represented in this form it is rarely used; consequently it is common practice to use complement arithmetic.

1.7 Complement arithmetic

This is a powerful yet simple technique which minimizes the hardware implementation of signed arithmetic operations in a digital machine. As a consequence of the use of complement arithmetic the process of subtraction becomes one of addition.

In the binary system there are two complements: (a) 2's complement or radix complement, and (b) 1's complement, or diminished radix complement. In practice, the use of 1's complement raises certain implementation difficulties and, as a consequence, signed arithmetic processes are performed in microprocessors using the 2's complement notation.

The 2's complement of a binary number X is defined by the equation

$$[X]_2 = 2^n - X$$

where n represents the number of binary digits contained in the number and X is the binary representation of the number. For $X = 1010$, $n = 4$, the 2's complement of this number can now be found using the above equation. Hence

$$
\begin{aligned}
[X]_2 &= 2^4 - 1010 \\
&= 10000 - 1010 \\
&= 0110
\end{aligned}
$$

There are two alternative methods available for determining the 2's complement of the number X. In the first method all the digits in the number are inverted and a 1 is added in to the least significant place, as illustrated below:

$$
\begin{array}{lll}
X & = & 1010 \\
& & 0101 \quad \text{Invert} \\
& & 1 \quad \text{Add} \\
\hline
[X]_2 & = & 0110
\end{array}
$$

For the second method, the lowest order 1 in X may be sensed and all succeeding higher order digits are inverted. For example

$$X = 1010$$

Invert   Sense
$$[X]_2 = 0110$$

In an 8-bit machine a positive number is represented by an 8-bit word where the most significant digit gives the sign of the number and the remaining seven bits represent its magnitude. Hence

$$+19 = 0,0010011$$

SD   Magnitude X

For the negative form of the same number the sign digit is 1 and the remaining seven digits are the number X represented in its 2's complement form, so that

$$-19 = 1,1101101$$

SD   2's complement $[X_2]$

The magnitude of a positive number in the 2's complement system is given by the equation

$$X = \sum_{i=0}^{i=n-2} X_i 2^i$$

while the magnitude of a negative number is given by

$$X = -2^{n-1} + \sum_{i=0}^{i=n-2} X_i 2^i$$

The combination of these two equations provides a general equation for the magnitude of a negative or positive number in the 2's complement form, and is given by

$$X = -(X_{n-1}2^{n-1}) + \sum_{i=0}^{i=n-2} X_i 2^i$$

Weighting value	SD -2^7 -128	2^6 64	2^5 32	2^4 16	2^3 8	2^2 4	2^1 2	2^0 1	Decimal value
	0	1	1	1	1	1	1	1	+127
	0	1	1	1	1	1	1	0	+126
	0	1	0	0	0	0	0	0	+64
	0	0	0	0	0	0	0	1	+1
	0	0	0	0	0	0	0	0	0
	1	1	1	1	1	1	1	1	−1
	1	1	1	1	1	1	1	0	−2
	1	0	0	0	0	0	0	0	−128

Figure 1.7 Tabular representation of 8-bit 2's complement numbers

The table shown in Figure 1.7 gives some of the 8-bit numbers available in 2's complement form and their corresponding decimal values. The range of these numbers is from +127 to −128 and it will be noted that it is not symmetrical since there is one more negative number available than there is positive.

1.8 Addition and subtraction of 2's complement numbers

Addition is carried out as if both the numbers are positive, and any carry from the sign bit is discarded. This will always give the correct result except when an arithmetic overflow has occurred. If the computer word is n bits long, an overflow is said to have occurred if the correct representation of the sum, including the sign bit, requires more than n bits.

Case I The addition of two 8-bit 2's complement numbers, both of which are positive and whose sum is $\leqslant 127$, is shown below:

-2^7	2^6	2^5	2^4	2^3	2^2	2^1	2^0	
0,	0	1	0	1	1	0	0	+ 44
0,	0	1	1	0	1	0	0	+ 52
0,	1	1	0	0	0	0	0	+ 96

The most significant digit is 0 and indicates a positive answer while the remaining seven digits give the magnitude of the answer, in this case 96, when expressed in decimal form.

Case II The addition of two positive numbers whose sum is >127 is illustrated below:

-2^7	2^6	2^5	2^4	2^3	2^2	2^1	2^0	
0,	1	1	0	0	0	0	1	+ 97
0,	0	1	1	0	0	0	0	+ 48
1,	0	0	1	0	0	0	1	+145

In this case, the most significant digit is 1 and indicates a negative answer. This is clearly wrong since both the augend and addend are positive, and the addition of two positive numbers requires a positive answer. In fact, an arithmetic overflow has occurred. The sum 145 cannot be represented by seven digits and an arithmetic overflow has occurred from the magnitude section of the number into the position occupied by the sign digit.

Case III The addition of positive and negative numbers where the positive number has the greater magnitude is illustrated in the example below:

	-2^7	2^6	2^5	2^4	2^3	2^2	2^1	2^0	
	0,	1	1	0	0	0	1	1	+ 99
	1,	1	0	1	1	1	1	1	− 33
Discard→(1)	0,	1	0	0	0	0	1	0	+ 66

In this case the answer is correct. However, there is a carry out of the sign bit which can be regarded as a register overflow. When implementing 2's complement arithmetic in a machine whose working register is eight bits wide, the register overflow is lost.

Case IV The addition of positive and negative numbers where the magnitude of the negative number is greater than that of the positive number is shown below:

-2^7	2^6	2^5	2^4	2^3	2^2	2^1	2^0	
0,	0	1	0	0	0	0	1	$+ 33$
1,	0	0	1	1	1	0	1	$- 99$
1,	0	1	1	1	1	1	0	$- 66$

Since the answer is negative it will appear in 2's complement form. In order to find the magnitude of the sum its 2's complement must be taken using the method described in Section 1.7 and repeated here:

-2^7	2^6	2^5	2^4	2^3	2^2	2^1	2^0	
1,	0	1	1	1	1	1	0	
0,	1	0	0	0	0	0	1	Invert
							1	Add
0,	1	0	0	0	0	1	0	$= 66$

Case V An example is shown here of the addition of two negative numbers whose |Sum|≤127:

	-2^7	2^6	2^5	2^4	2^3	2^2	2^1	2^0	
	1,	1	1	0	0	0	1	1	$- 29$
	1,	1	1	0	0	0	0	0	$- 32$
Discard→(1)	1,	1	0	0	0	0	1	1	$- 61$

The answer is negative, as indicated by the fact that the most significant digit is 1. A carry is generated out of the sign digit and this register overflow has to be discarded. As in Case IV, in order to obtain the magnitude of the sum its 2's complement must be taken.

Case VI The addition of two negative numbers when the |Sum|≥127 is illustrated below:

	-2^7	2^6	2^5	2^4	2^3	2^2	2^1	2^0	
	1,	1	0	1	0	1	1	1	$- 41$
	1,	0	1	0	0	0	0	1	$- 95$
Discard→(1)	0,	1	1	1	1	0	0	0	-136

If the register overflow is discarded the answer would be interpreted as +120, which is obviously incorrect. The reason for this error is that −136 cannot be represented by seven binary digits. The minimum negative number that can be represented by this number of digits is −128.

1.9 Multiplication of 2's complement numbers

Multiplication of two signed binary numbers using the shift and add principle described in Section 1.5 leads to a number of complications. Direct multiplication requires that the sign of the multiplier has to be tested and if it is negative a correction has to be made to the product.

An alternative method of approach saves the signs of the multiplier and the multiplicand. Negative numbers are then converted to positive numbers and the multiplication is performed using the method of Section 1.5. If the signs of the multiplier and multiplicand are different the answer is negative and the product has to be converted back to the 2's complement form.

The complications described above can be avoided by the use of Booth's method, where the procedure is the same irrespective of the sign of the multiplier. In this method the multiplier digits are examined in pairs starting at the least significant end, and the multiplicand is then added or subtracted from the accumulating partial product according to the information obtained from the comparison of the multiplier bits.

The algorithm for the method may be stated as follows:

(1) If the compared digits are 00 or 11 do nothing, and shift the multiplier and partial product one place right.
(2) If the compared digits are 10, subtract the multiplicand from the accumulated partial product and shift right one place.
(3) If the compared digits are 01, add the multiplicand to the accumulated partial product and shift right one place.

When the number to be shifted right is negative, the digit moved into the most significant position is a 1, and when it is positive, the digit moved into the most significant position is 0.

An example of the method is shown below, where the multiplier is a negative number and the multiplicand is a positive number:

Multiplicand	0,1100	+12
Multiplier	1,0010[0]	−14
	0,000000000	
00 Shift right	0,000000000	1st partial product
10 Subtract m'cand	1,0100	
	1,010000000	2nd partial product
Shift Right	1,101000000	
01 Add m'cand	0,1100	
Discard carry (1)	0,011000000	3rd partial product

Shift right	0,001100000
00 Shift right	0,000110000 4th partial product
10 Subtract m'cand	1,0100

1,010110000

Shift right	1,101011000 Negative product
	0,010100111 Invert
	1 Add

0,010101000 Magnitude = $\overline{168}$

It will be noticed that for the first comparison an assumed 0 is placed behind the least significant digit, and the first comparison is made between this assumed digit and the least significant digit.

There are clearly three other possibilities:

(1) Multiplicand (+)   Multiplier (+)
(2) Multiplicand (−)   Multiplier (+)
(3) Multiplicand (−)   Multiplier (−)

It is left to the reader to verify Booth's method for these three cases.

1.10 Division of 2's complement numbers

The process of binary division is illustrated below for two signed binary numbers represented in 2's complement form. Initially, the signs of the dividend and the divisor have to be examined to determine the signs of the quotient and the remainder. The magnitude of the quotient and the remainder are then determined by division after converting both the divisor and dividend into positive numbers, as shown in the following example. When the magnitudes of the quotient and the remainder have been determined they are then converted into their correct 2's complement representation.

The method used in the following example is the *restoring algorithm*. The principle of the method is that an attempt is made to subtract the divisor from the dividend. If the result of the subtraction is negative the quotient bit is zero and the dividend has to be restored. If the result after subtraction is positive, the quotient bit is 1 and the subtraction is valid. In either case the divisor is shifted right relative to the dividend and the procedure is repeated. Since for the subtraction process the divisor is negative it is represented in 2's complement form. Consequently, when the shift right of the divisor occurs the digit moved into the most significant position will always be a 1.

Divisor = +12   = 0,1100
2's complement of divisor   = 1,0100
Dividend = +179   = 0,10110011

```
                  0,1110
                 ─────────
                 0,10110011
                 1,0100        Subtract divisor
                 ─────────
                 1,11110011    Answer negative, restore
                               dividend, Q=0
                 0,10110011    Dividend restored
                 1,10100       Shift divisor right and subtract
                 ─────────
Discard   (1)    0,01010011    Answer positive, Q=1
carry            1,110100      Shift divisor right and subtract
                 ─────────
Discard   (1)    0,00100011    Answer positive, Q=1
carry            1,1110100     Shift divisor right and subtract
                 ─────────
Discard   (1)    0,00001011    Answer positive, Q=1
carry            1,11110100    Shift divisor right and subtract
                 ─────────
                 1,11111111    Answer negative, restore
                               dividend, Q=0
Remainder        0,00001011    Dividend restored
```

Answer: Quotient 0,1110 = 14
        Remainder 0,00001011 = 11

The algorithm used to perform the division process in the above example can be summarized as follows:

(1) Align the most significant bits of the divisor and dividend.
(2) Add the 2's complement of the divisor to the dividend.
(3) If now the most significant digit is 1, the answer is negative, so restore the dividend, shift the divisor right and record the quotient bit Q =0. *Or*
(4) If the most significant digit is 0 the answer is positive, so shift the divisor right and record the quotient bit Q = 1.
(5) Repeat (2), (3) and (4) until the least significant digits of the divisor and dividend are aligned.

1.11 Binary coded decimal (BCD) arithmetic

The decimal digits from 0 to 9 inclusive can each be represented by four binary digits as illustrated in Figure 1.8. Only ten out of the 16 possible four-bit combinations are used and consequently there are six invalid code combinations. They are 1010, 1011, 1100, 1101, 1110 and 1111.

d	A	B	C	D
0	0	0	0	0
1	0	0	0	1
2	0	0	1	0
3	0	0	1	1
4	0	1	0	0
5	0	1	0	1
6	0	1	1	0
7	0	1	1	1
8	1	0	0	0
9	1	0	0	1

Figure 1.8 the BCD code for decimal digits

A decimal readout is a frequent requirement of a microcomputer system. Seven-segment displays that can be driven by BCD-to-seven-segment decoder chips are often used to give a visual decimal display. In some cases both the input and output data are in decimal form and, providing only a limited amount of computation is required in the machine, this can be performed using BCD arithmetic. If, however, considerable calculation is to be performed on the input data, then it is advisable to convert to a pure binary representation, do the necessary computation, and then reconvert to BCD form and output.

When two unsigned BCD numbers are added together, incorrect answers may be obtained, as can be seen by an examination of the following examples.

Case I   S ≤ 9

$$
\begin{array}{rl}
4 & 0100 \\
+5 & 0101 \\
\hline
+9 & 1001
\end{array}
$$

Direct addition of the codes in this case gives the correct answer.

Case II   9 < S ≤ 15

$$
\begin{array}{rl}
7 & 0111 \\
+6 & 0110 \\
\hline
+13 & 1101
\end{array}
$$

In this case direct addition produces an invalid code. The answer should be

13 = 0001   0011

and eight bits are required to represent the sum.

Case III   15 < S ≤ 19   9   1001
                                +8   1000
                                ─────────────
                               +17   1,0001

Direct addition generates a carry and produces a valid but incorrect code. The answer should be

17 = 0001   0111

In both Cases II and III, corrections have to be made before the correct sum is obtained. For the case of 9 < S ≤ 15, the sum obtained 13 = 1101 is 10 too many in the least significant four bits. The correct answer is obtained by subtracting 10 but, rather than subtract, it is more convenient to add the 2's complement of 10, as shown below:

2's complement of 10 = 0110

    7   0111   Augend
   +6   0110   Addend
         ─────────
          1101
          0110   Add 2's complement of 10
   ───────────────
  +13   1,0011   Sum

After the correction is made, a carry is generated and the correct value for the least significant decimal digit is obtained.

In the same way the correction can be applied to the example shown under the heading Case III, and the correct answer is then obtained as illustrated below:

    9   1001   Augend
   +8   1000   Addend
         ─────────
         1,0001
          0110   Add 2's complement of 10
   ───────────────
   17   1,0111   Sum

Since the word length in an 8-bit microprocessor is eight bits, it can be used to represent two decimal digits. An example of the addition of two 8-bit BCD words is shown below:

   49   0100   1001   Augend
   33   0011   0011   Addend
        ──────────────
         0111   1100
         0000   0110   BCD correction
        ──────────────
         0111   ⟋0010
            1 ⟋   0   Carry
   ──────────────────────
   82   1000   0010   Sum

1.12 10's complement arithmetic

The 10's complement of a number in the decimal system is analogous to the 2's complement in the binary system and is defined by the equation

$$[X]_{10} = 10^n - X$$

where n = the number of decimal digits contained in the decimal number X and $[X]_{10}$ represents its 10's complement. For $X = 823$, $n = 3$ and

$$[X]_{10} = 10^3 - 823 = 177$$

An alternative method of finding the 10's complement is to subtract each decimal digit in the number from 9 and then add on 1 in the least significant place, as shown in the following example:

$$
\begin{array}{r}
999 \\
823 \\
\hline
176 \\
1 \\
\hline
177
\end{array}
$$

Subtraction can now be carried out as an addition using the 10's complement form for negative numbers. Two cases will be considered.

Case I Addition of positive and negative numbers where the positive number has the greatest magnitude. For example, $62 - 55$ can be performed by taking the 10's complement of 55 and adding it to 62, as shown below:

10's complement of $55 = 45 = 0100 \quad 0101$

	62	0110	0010	Minuend
−55	0100	0101	Subtrahend

07	1010	0111	
	0110	0000	BCD correction

Discard   →(1) 0000   0111   Difference

Case II Addition of positive and negative numbers, the negative number having the greater magnitude. In this case the subtraction to be carried out is $55-62$ and it can be performed by finding the 10's complement of 62 and adding it to 55, as shown below:

10's complement of $62 = 38 = 0011 \quad 1000$

55	0101	0101	Minuend
−62	0011	1000	Subtrahend
−07	1000	1101	
	0000	0110	BCD correction
	1		Carry
	1001	0011	Difference

Since the answer must be negative, it is expressed in 10's complement form. In order to obtain the magnitude it is necessary to take the 10's complement of the answer 93 which is the required magnitude 07.

It should be noticed that when the difference is positive as in Case I, a carry is generated in the most significant place of the addition, and this can be used to indicate the positive nature of the answer. In Case II, where the answer is negative, no carry is generated in the most significant place of the addition. The generation of this carry can be used to distinguish between positive and negative answers.

1.13 Signed BCD arithmetic

For signed binary coded decimal arithmetic, the sign of a number is represented by four binary digits. In the case of a positive number the sign digit is 0 and is represented by the four binary digits 0000, while for a negative number the sign digit is 9, which is represented by the four binary digits 1001. Hence a positive number such as +62 would be represented in signed BCD form as

$+62 = 062 = 0000 \quad 0110 \quad 0010$

and a negative number such as −55 would be represented by its 10's complement and be preceded by the sign digit, hence

$-55 = 945 = 1001 \quad 0100 \quad 0101$

The subtraction of 55 from 62 using 10's complement arithmetic becomes an addition process and it is left to the reader to perform this operation using the techniques previously described.

The range of decimal numbers available in a two-byte BCD representation is shown in Figure 1.9.

1.14 Multiplication and division of BCD numbers

Multiplication and division of BCD numbers is not normally performed in a microprocessor. This is because the techniques

Decimal no.	10's complement code
+999	0999
⋮	⋮
+2	0002
+1	0001
0	0000
−1	9999
−2	9998
⋮	⋮
−999	9001
−1000	9000

Figure 1.9 The range of decimal numbers available in a two-byte BCD representation

used are very lengthy and consequently machine execution time is long. A more practical approach is to convert the binary coded decimal numbers to pure binary, perform the multiplication or division and, after its completion, reconvert to BCD.

1.15 Decimal to binary conversion

As just explained, multiplication or division of BCD numbers is not normally carried out in small computing machines like microprocessors. Consequently, it is a frequent requirement that decimal data, represented in BCD form, has to be converted into pure binary so that the required computation can be carried out using binary arithmetic.

There are a variety of ways of implementing BCD to binary conversion. One method used is based on the equation presented in Section 1.2 of this chapter for the decimal magnitude $(N)_{10}$ of a number in any base system. In the context of decimal to binary conversion the equation is rewritten below specifically for the decimal system:

$$(N)_{10} = d_{n-1}10^{n-1} + d_{n-2}10^{n-2} + \ldots + d_2 10^2 + d_1 10^1 + d_0 10^0$$

where n is the number of decimal digits and d is the decimal positional coefficient. This equation can be manipulated into the nested form shown below:

$$(N)_{10} = (----((d_{n-1})10 + d_{n-2})10 + d_{n-3})10 --- d_2)10 + d_1)10 + d_0$$

Assuming that a two decimal digit number is to be converted, then the relevant portion of this equation is

$$(N)_{10} = 10d_1 + d_0$$

where d_1 and d_0 represent the two decimal digits. After further modification, the equation becomes

$$\begin{aligned}(N)_{10} &= 8d_1 + 2d_1 + d_0\\ &= 2^3d_1 + 2d_1 + d_0\end{aligned}$$

This equation can be used for developing a method for converting a number such as 39 = 0011   1001 to pure binary. For this number $d_1 = 0011$ and $d_0 = 1001$.

Multiplying d_1 by 2 is equivalent to shifting the binary digits left one place; for example

$$\begin{aligned}d_1 &= 0011 = 3\\ 2d_1 &= 0110 = 6\end{aligned}$$

and it follows that multiplying d_1 by 2^3 is equivalent to shifting the binary digits three places to the left. Hence

$$\begin{aligned}2^3d_1 &= 11000 = 24\\ \text{Then}\quad (39)_{10} &= \begin{array}{ll}11000 & 8d_1\\ 0110 & 2d_1\\ 1001 & d_0\end{array}\\ &\quad\overline{100111}\end{aligned}$$

1.16 Fractional binary numbers

An 8-bit unsigned fraction can be represented by an 8-bit word in a digital machine simply by moving the binary point eight places to the left. The binary point is then imagined to occupy a position just to the left of the most significant digit of the 8-bit word.

Shifting the binary point in this manner is equivalent to dividing the integral number represented by the 8-bit word by 256. Hence the 8-bit unsigned fractional number can be represented by the following equation

$$X = .X_7X_6 \text{------} X_0 \quad \begin{array}{l}\text{Binary point shifted}\\ \text{8 places left}\end{array}$$

The decimal magnitude of the fractional number represented by this equation is

$$(N_f)_{10} = X_7 2^{-1} + X_6 2^{-2} + X_5 2^{-3} + \text{-----} X_0 2^{-8}$$

If the coefficients X_7 to X_0 inclusive in this equation are all equal to 0 then

$(N_f)_{10}$ = .00000000 = 0/256 = 0

Alternatively, if X_7 to X_0 inclusive are all equal to 1

$(N_f)_{10}$ = .11111111 = 255/256 = 0.99609375

When the machine is carrying out signed fractional arithmetic, the most significant digit X_7 is the sign digit and the binary point is then imagined to lie between X_7 and X_6.

The magnitude of either a positive or negative fractional number in the 2's complement system is given by the equation

$$X = (-X_{n-1}) + \sum_{i=0}^{n-2} X_i\, 2^{-[n-(i+1)]}$$

The table shown in Figure 1.10 gives some of the 8-bit fractional numbers available in 2's complement form and their corresponding decimal values. The range of numbers available is from +0.99218750 to −1 and, as in the case of integral numbers, the range is not symmetrical.

Weighting value	SD −1	2^{-1} $\frac{1}{2}$	2^{-2} $\frac{1}{4}$	2^{-3} $\frac{1}{8}$	2^{-4} $\frac{1}{16}$	2^{-5} $\frac{1}{32}$	2^{-6} $\frac{1}{64}$	2^{-7} $\frac{1}{128}$	Decimal value
	0	1	1	1	1	1	1	1	+.99218750
	0	1	1	1	1	1	1	0	+.98437500
	0	1	0	0	0	0	0	0	+.50000000
	0	0	0	0	0	0	0	1	+.00781250
	0	0	0	0	0	0	0	0	0
	−1	1	1	1	1	1	1	1	−.00781250
	−1	1	1	1	1	1	1	0	−.01562500
	−1	0	0	0	0	0	0	0	−1.00000000

Figure 1.10 Tabular representation of 8-bit fractional numbers

An integral decimal number can be exactly represented in binary form providing there are sufficient binary digits available. For example, any decimal number ≤255 can be exactly represented by eight binary digits. Therefore, providing the decimal range of the eight binary digits is not exceeded, an exact representation of the number is available.

However, in a machine having a computer word length of eight bits, fractional representation of a decimal number is only exact in those cases where it can be represented by eight or less bits. For example,

$$(0.375)_{10} = (0.01100000)_2$$

and needs a maximum of three binary digits for exact representation. However,

$$(0.265)_{10} = (0.01000011)_2$$

has no exact representation in eight binary digits.

The approximate representation shown above has an error which is less than $2^{-8} = 1/256$ and is obtained by restricting the binary representation to eight digits. This type of error is called a *truncation error*.

Alternatively, $(0.265)_{10}$ can be represented by a nine-digit fractional binary number, so that

$$(0.265)_{10} = (0.010000111)_2$$

Another method of approximation then involves rounding to the nearest representable value containing eight digits, namely

$$(0.265)_{10} = (0.01000100)_2$$

Rounding in this way provides a more accurate approximation with the *rounding error* being $<2^{-(n+1)}$ for an n-digit representation.

Clearly, arithmetic operations involving fractions will normally lead to error, even when the fraction is exactly representable. The error occurs because, after an arithmetic operation such as multiplication, the result has to be either truncated or rounded. If either of these methods of approximation produces an unacceptable error then it will be necessary to employ more binary digits to represent the original numbers.

Problems

1.1   Find the decimal equivalent of the following numbers:
(a) $(10101101)_2$ (b) $(247)_8$ (c) $(4FC3)_{16}$

1.2   Convert $(242)_{10}$ into (a) binary (b) octal and (c) hexadecimal

1.3   Convert the fractional number $(0.87)_{10}$ into (a) binary (b) octal and (c) hexadecimal

1.4   Determine the binary equivalent of:
      (a) $(232)_4$ (b) $(232)_8$ and (c) $(AB8)_{16}$

1.5   Determine (a) the base 4 (b) octal and (c) hexadecimal
      equivalent of:
      $(1010111000011000)_2$

1.6   If $(25)_{10} = (34)_b$ and $(728)_{10} = (1330)_b$, find b in both cases.

1.7   The arithmetic operations given are correct in one number
      system. Determine the possible number bases for each
      operation:

      (a) $\dfrac{41}{3} = 13$

      (b) $24 + 32 + 12 + 40 = 213$

      (c) $\dfrac{48}{4} = 12$

      (d) $(51)^{\frac{1}{2}} = 6$

1.8   Given the binary numbers

      $p = 10101100$ $q = 01011010$ and $r = 00100111$

      perform the following operations in pure binary:

      (a) $p + q$ (c) $p \times q$
      (b) $p - r$ (d) $p \div r$

1.9   Express the numbers $(-29)_{10}$ and $(+82)_{10}$ as 8-bit binary
      numbers in the following forms:

      (a) sign and magnitude,
      (b) 2's complement, and
      (c) 1's complement.

1.10 Using 2's complement arithmetic, and representing all
      numbers in terms of eight binary digits where the most
      significant digit is the sign digit, perform the following
      arithmetic operations and comment on your answers:

+ 82	+79	−79	−29
+ 63	−33	+33	−37
+145	+46	−46	−66

      Repeat the above operations using 1's complement arith-
      metic.

1.11 Using Booth's Algorithm, perform the following calculations:

   (a) -14×-21
   (b) $-7 \times +17$

1.12 Using the restoring algorithm, perform the following calculation:

   $273 \div 19$

1.13 Express $(82)_{10}$ and $(29)_{10}$ in the $8-4-2-1$ binary coded decimal form. Add the two numbers together in the BCD form, and design a logic circuit for performing this operation using a 4-bit adder as a basic building block.

1.14 Determine the 10's and 9's complement of the decimal numbers 229 and 302 and express them in BCD form. Perform the BCD arithmetic operation $302 - 229$ using signed 10's complement arithmetic and the operation $229 - 302$ using signed 9's complement arithmetic.

2 Logic

2.1 Introduction

In addition to its ability to perform arithmetic operations, a microprocessor can be programmed to implement logic operations such as AND, OR, Exclusive-OR(XOR) and NOT. Both the arithmetic and the logic operations are performed in the arithmetic/logic unit (ALU) which is a component part of the microprocessor. Additionally, logic elements are used as component parts of microcomputer systems. They may be discrete elements such as gates or flip-flops, or, alternatively, they may be MSI chips such as multiplexers, decoders, encoders, etc. The purpose of this chapter is to present a survey of the behaviour of the most commonly used logical elements that the engineer is likely to encounter in a microcomputer system.

2.2 Gates

The six most commonly used gates are the AND, OR, NAND, NOR, XOR and the Exclusive-NOR gates. A composite truth table for these six gates is shown in Figure 2.1(a). It is assumed that in each case the gate has two inputs A and B, and the conventional diagrams for these gates are shown in Figure 2.1(b). The Boolean functions implemented by each of the gates are shown adjacent to their output lines.

The small circles appearing at the gate end of the output lines of the NAND, NOR and Exclusive-NOR gates are called negation circles, and they indicate, for example, that a NAND gate can be regarded as an AND gate whose output has been negated (inverted).

A	B	AND	OR	NAND	NOR	XOR	Exc–NOR
0	0	0	0	1	1	0	1
0	1	0	1	1	0	1	0
1	0	0	1	1	0	1	0
1	1	1	1	0	0	0	1

(a)

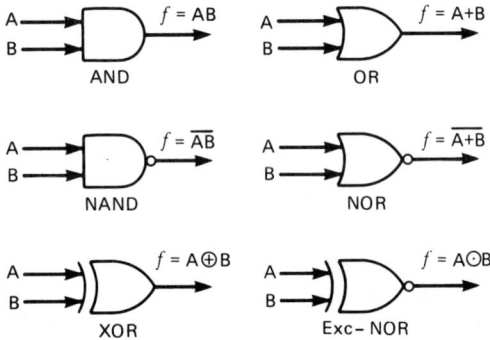

(b)

Figure 2.1 (a) Composite truth table for the AND, OR, NAND, NOR, XOR and Exclusive-NOR functions (b) Conventional gate representations

2.3 Inverters

The inversion of a Boolean function or a Boolean variable can be implemented by an inverter whose conventional diagrammatic representation is shown in Figure 2.2(a) along with its truth table in Figure 2.2(b).

It is also possible to use both NAND and NOR gates as inverters by a simple modification of the input connections. In the case of a two-input NAND gate, one of the two input leads is held permanently high while the other input lead carries the variable to be inverted, as shown in Figure 2.3(a). For a two-input NOR gate,

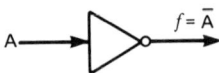

A	f
0	1
1	0

(a)   (b)

Figure 2.2 (a) Conventional representation of an inverter (b) Truth table one input lead is permanently held low by grounding it, while the remaining lead carries the variable to be inverted, as shown in Figure 2.3(b). It is worth noting at this point that unused gate inputs should never be left disconnected as in the case of the inverter circuits just described, they should be connected to either logical 1 or logical 0.

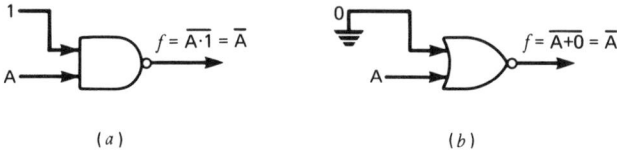

(a)   (b)

Figure 2.3 (a) A two-input NAND gate connected as an inverter (b) A two-input NOR gate connected as an inverter

2.4 Alternative logic representations

An alternative representation of the AND and OR gates can be obtained by the application of de Morgan's theorem to the two function equations. For example, the equation of the NAND function is

$$f = \overline{AB}$$

but, by applying de Morgan's theorem, it can be written as

$$f = \overline{AB} = \overline{A} + \overline{B}$$

In other words, the NAND function can be implemented by an OR gate having inverted inputs for the variables A and B. The two possible implementations of the NAND function are shown in Figure 2.4(a). The negation circles at the inputs of the OR gate indicate the inversion of the input variables A and B.

Similarly, applying de Morgan's theorem to the NOR function

$$f = \overline{A + B}$$

it can be written as

$$f = \overline{A + B} = \overline{A}\,\overline{B}$$

Hence, the OR function can be implemented by an AND gate having inverted inputs for the variables A and B, and the two implementations are shown in Figure 2.4(b).

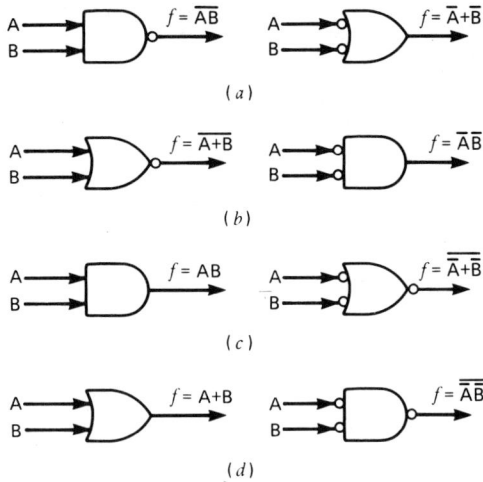

Figure 2.4 The two implementations of (a) the NAND function (b) the NOR function (c) the OR function

Equivalent implementations for the AND and OR functions can be obtained by inverting both sides of the NAND and NOR equations. Therefore

$$AB = \overline{A} + \overline{B}$$

and

$$A + B = \overline{A}\ \overline{B}$$

The alternative implementation of the AND function is a NOR gate having inverted inputs for the variables A and B, and for the OR function it is a NAND gate again having inverted inputs for the variables A and B. These alternative implementations are illustrated in Figures 2.4(c) and (d).

Either of the alternative gate representations for the four logic functions shown in Figure 2.4 can be used on a logic diagram. However, the two gate representations of a particular logic function can have different interpretations. For example, the behaviour of the NAND gate shown in Figure 2.5(a) can be interpreted by saying that the output goes low when all the inputs are high, while for the alternative representation of the NAND gate shown in Figure 2.5(b), an interpretation of its behaviour is that the output goes high when any input is low.

The representation shown in Figure 2.5(*a*) would be used when the normal output is high, but when the gate is activated its output goes low. This gate representation of the NAND function is said to have an active low output. Alternatively, the representation shown in Figure 2.5(*b*) would be used when the normal output is low but, when activated, the gate output goes high. This gate representation of the NAND function is said to have an active high output.

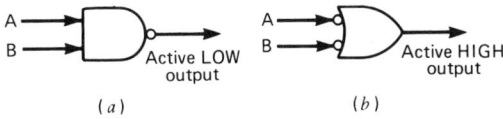

Figure 2.5 (*a*) NAND gate, active low output (*b*) NAND gate, active high output

The two alternative representations of each of the three remaining logic functions shown in Figures 2.4(*b*), (*c*) and (*d*) can be interpreted in a similar fashion. Those gates with a negation circle at the output are defined as having an active low output, while those without a negation circle are defined as having an active high output.

2.5 The tri-state gate

Another important device used in microprocessor systems is the tri-state gate, of which there are inverting and non-inverting types. Illustrations of the two types with their associated truth tables are shown in Figures 2.6(*a*) and (*b*).

The E line can be regarded as an enable line which, when activated, will allow either the transmission or inversion of the data input A. When the enable line is not activated the gate presents a high output impedance and transmission of true or inverted data is inhibited.

Many tri-state devices are used in microprocessor systems because of the high impedance property they possess when disabled. For example, the data lines leaving a particular integrated circuit device in a microprocessor system are normally tri-stated so that the device can be effectively isolated from the system data bus. A diagram illustrating the arrangement is shown in Figure 2.7. Data from the device cannot be transferred to the system interconnecting bus until each of the tri-state gates has been enabled by the signal E.

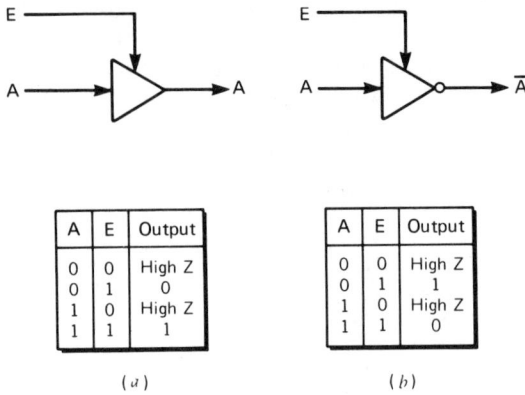

A	E	Output
0	0	High Z
0	1	0
1	0	High Z
1	1	1

A	E	Output
0	0	High Z
0	1	1
1	0	High Z
1	1	0

(a)   (b)

Figure 2.6 (a) The tri-state gate and its truth table (b) The inverting tri-state gate and its truth table

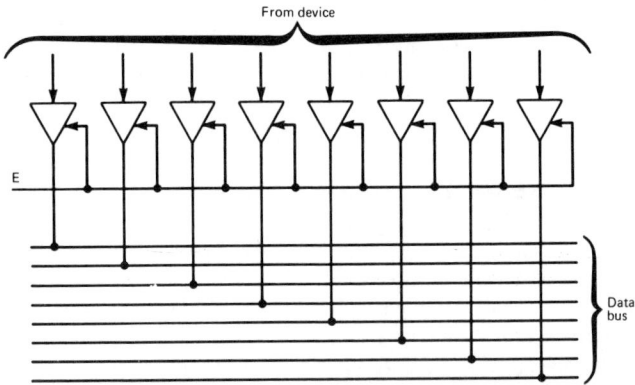

Figure 2.7 Device connection to system data bus via tri-state buffers

In certain circumstances data may have to be transmitted from the device to the interconnecting bus and also received by the device from the same bus. This will require a bidirectional capability, as illustrated in Figure 2.8. The transmission and receipt of data is controlled by two selection signals, E_1 and E_2. When $E_1 = 1$ and $E_2 = 0$, data is transmitted from the device to the bus, and when $E_1 = 0$ and $E_2 = 1$ data can be received from the bus. When $E_1 = E_2 = 0$, both the device and the bus are tri-stated and data can pass in neither direction.

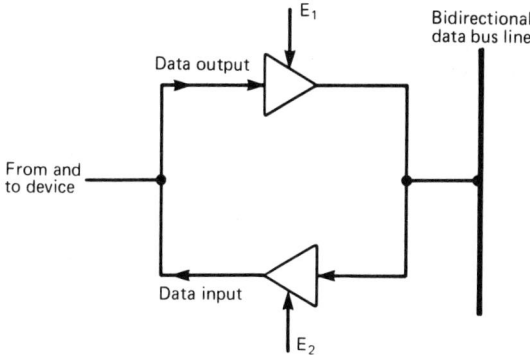

Figure 2.8 Tri-state connection to bidirectional data bus

2.6 The D-type flip-flop

A single-bit memory element is a device that will store one binary digit. The most widely used device of this kind in a microprocessor system is a D-type flip-flop, although it should be noted that other types of flip-flops such as T, SR and JK are also widely available and are sometimes used in microprocessor systems. A D-type flip-flop is a bistable element, which implies that it has two stable states, and conventionally it can be represented by the diagram shown in Figure 2.9(a).

For this D-type flip-flop there is one input or data line D and two complementary output lines, Q and $\bar{Q}$. The two stable states are defined as (a) Q = 0 and $\bar{Q}$ = 1, and (b) Q = 1 and $\bar{Q}$ = 0. Control of the transfer of data D to the output Q line is achieved via the clock line.

The logical behaviour of the D-type flip-flop is defined by the state table shown in Figure 2.9(b). In the first two columns of this

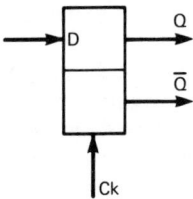

Present state		Next state
D^t	Q^t	$Q^{t+\delta t}$
0	0	0
0	1	0
1	0	1
1	1	1

(a)   (b)

Figure 2.9 (a) The D-type flip-flop (b) Truth table

table every possible combination of the present input D^t and the present output Q^t are tabulated. The last column gives the resulting next state of the flip-flop $Q^{t+\delta t}$.

From the state table the following equation is obtained:

$$Q^{t+\delta t} = (D\bar{Q} + DQ)^t$$

or

$$Q^{t+\delta t} = D^t$$

This is called the characteristic equation of the flip-flop and it indicates that the next state of the output $Q^{t+\delta t}$ is identical to the present state of the input data D^t.

2.7 The edge-triggered D-type flip-flop

This device can be regarded as one that transfers data at the D-input to the Q-output. Such a transfer is initiated by a clock signal, which may be regarded as an enabling signal that allows the transfer to take place. A typical example of a clock signal is shown in Figure 2.10. Transfer of data normally takes place on either the leading or trailing edge of a clock pulse. It should be noticed that there can be no change of the Q output when the clock is low. This period is referred to as the asynchronous period of the clock signal and, in effect, during this period the flip-flop is disabled.

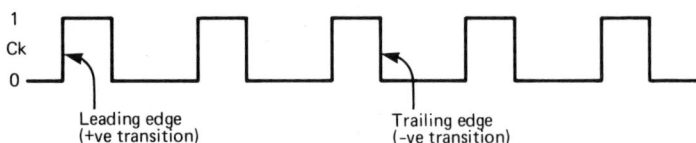

Leading edge
(+ve transition)

Trailing edge
(−ve transition)

Figure 2.10 Typical clock waveform

A flip-flop that changes its state on a clock transition is called an *edge-triggered flip-flop*.

The sensitivity of the flip-flop to a clock transition is indicated by $>$ at the clock input. If it is sensitive to a negative-going transition of the clock then, additionally, an inversion circle is used in conjunction with the previously defined symbol for edge triggering. Conventional representations of the negative and positive edge-triggered D-type flip-flops are shown in Figures 2.11(*a*) and (*b*), together with timing diagrams that show the variation of Q with time for a given input waveform D.

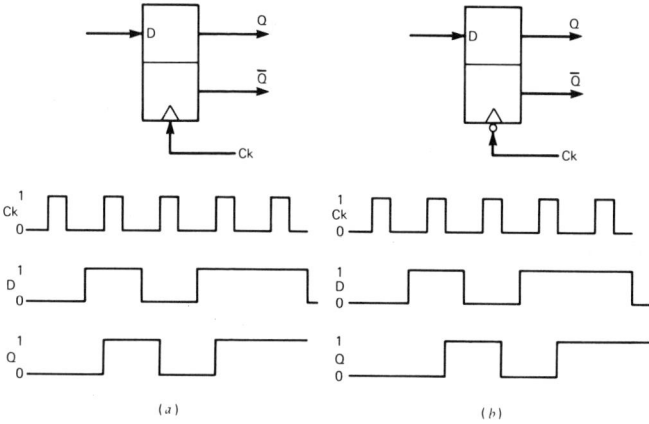

Figure 2.11 (a) The positive edge triggered D-type flip-flop and its timing diagram
(b) The negative edge triggered D-type flip-flop and its timing diagram

2.8 The level-triggered D-type flip-flop

An alternative form of the D-type flip-flop is the *latch* or
level-triggered flip-flop. It is similar to the edge-triggered flip-flop
except that the output Q changes with the input D while the clock
is high. When the clock makes a transition from 1 to 0 the data
present at the D-input is latched. This kind of D-type flip-flop is
said to be positive level triggered. Alternatively, if the output Q

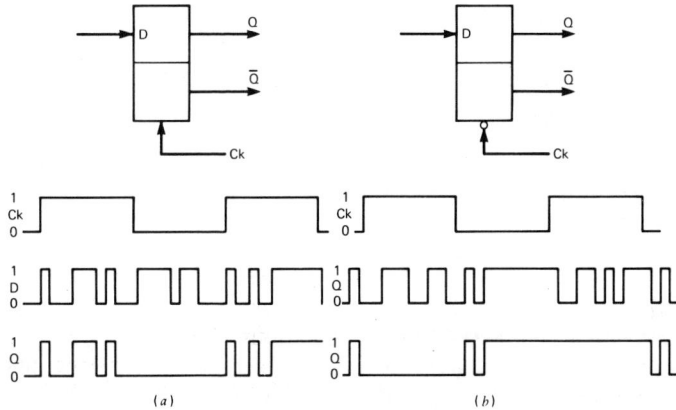

Figure 2.12 (a) The positive level triggered D-type flip-flop and its timing diagram
(b) The negative level triggered D-type flip-flop and its timing diagram

follows the input D when the clock is low and latching takes place on the positive-going transition of the clock, the flip-flop is said to be negative level triggered. Conventional representations of the positive and negative level triggered D-type flip-flops are shown in Figures 2.12(a) and (b) with their associated timing diagrams showing the variation of Q, the output, with the input D.

2.9 Asynchronous controls

There are two additional inputs commonly available on most flip-flops. These are termed the asynchronous controls Preset and Clear and are employed when the clock signal is low, that is when the flip-flop is disabled as far as its D-input is concerned. A conventional representation of a D-type flip-flop incorporating Preset and Clear is shown in Figure 2.13(a). The inversion signals on the preset and clear lines indicate that these two signals are active low. Operation of these controls is described by the truth table shown in Figure 2.13(b). Note that Pr and Cl being simultaneously equal to logical 0 is a forbidden combination.

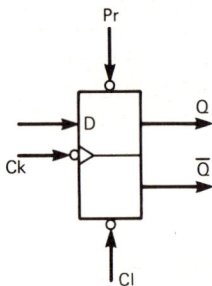

	Ck	Pr	Cl	Q	
FF enabled	1	1	1	*	Normal operation
Preset	0	0	1	1	
Clear	0	1	0	0	
	X	0	0		Forbidden combination

(a) (b)

Figure 2.13 (a) Asynchronous control, Preset and Clear applied to an edge-triggered D-type flip-flop (b) Table describing the behaviour of the Preset and Clear controls

2.10 Flip-flop conversion

If it is required, it is a simple matter to convert both JK and SR flip-flops into D-type flip-flops by connecting an inverter between the two input lines, as indicated in Figures 2.14(a) and (b).

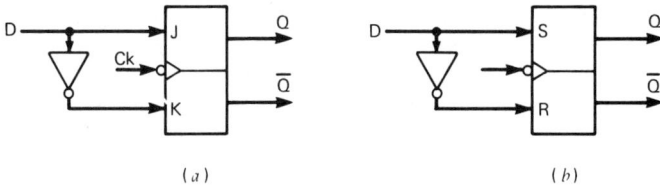

Figure 2.14 (a) Conversion of a JK flip-flop to a D-type flip-flop (b) Conversion of an SR flip-flop to a D-type flip-flop

2.11 Flip-flop operating characteristics

Certain timing constraints have to be observed to ensure correct flip-flop operation. For example, the input data must be stable for a minimum period t_{su} before a clock transition can be allowed to occur. This time t_{su} is called the *set-up time* and normally lies within the range 5 to 50 ns.

Additionally, the input data must be stable for a minimum period t_h after the clock transition. This time t_h is called the *hold time* and is in the range 0 to 10 ns. For the case of $t_h = 0$ the input signal is terminated at the same instant the clock transition occurs. Hold and set-up times are illustrated in Figure 2.15. Notice that these times are measured from the 50 per cent amplitude points of the clock and input waveforms.

Manufacturers also specify a minimum clock pulse width t_w (see Figure 2.15) and a maximum clock frequency f_m. For frequencies above f_m the flip-flop would not be able to respond quickly enough to the rapidly occurring clock transitions.

There is, of course, always a delay occurring between a clock transition and a change in output level. This is called the *propagation delay* of the flip-flop. There are two values for the

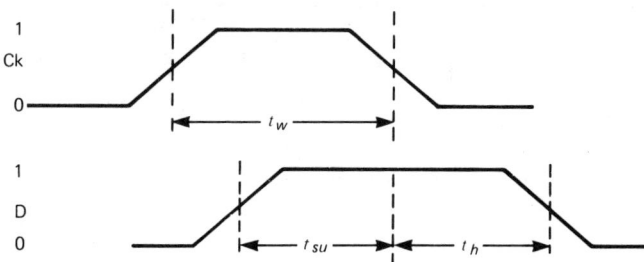

Figure 2.15 Timing diagram illustrating set-up time t_{su}, hold time t_h, and minimum allowable clock width t_w

propagation delay: one when Q is changing from low to high, t_{plh}; and the other when Q changes from high to low, t_{phl}. Values for these delays are quoted on the manufacturers' data sheets. It is also common practice for the manufacturers to specify the delay occurring between the application of either a preset or clear, and the resulting change of flip-flop state.

2.12 Registers

A fundamental component found in any microcomputer system is the register. The microprocessor chip itself contains a number of registers such as the accumulator and the instruction register. Additionally, registers are found elsewhere in the microcomputer system; for example, a memory chip may be regarded as an array of addressable registers, and an input/output port can also be regarded as a register.

A register consists of an array of flip-flops and is used for storing binary information on either a permanent or temporary basis. The period of retention may be as short as nanoseconds or, as in the case of a read only memory (ROM), indefinite.

A block diagram of a register is shown in Figure 2.16. Data is entered into the register by means of the WRITE control signal and will remain there until fresh data is written into the register or until it is cleared. When required, data can be read from the

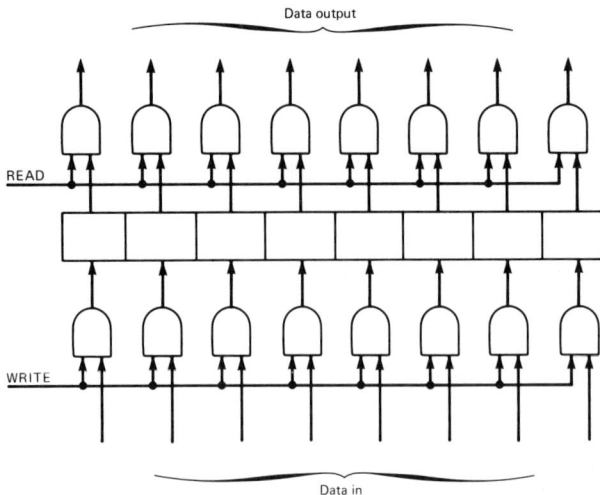

Figure 2.16 Basic parallel in/parallel out register

register by the application of the appropriate logic level to the READ control line. In this example, data is moved in and out of the register in parallel form, and for this reason it is called a parallel in/parallel out type of register, or PIPO.

2.13 Data transfer operations

A typical operation performed in a microprocessor is the parallel transfer of data from one register to another. The process is illustrated in Figure 2.17.

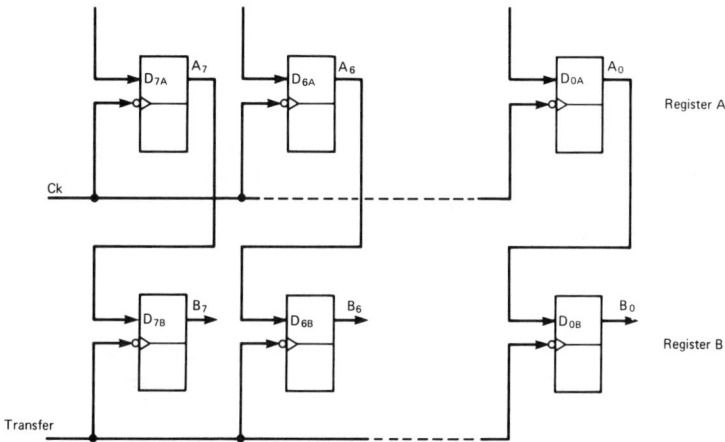

Figure 2.17 Parallel transfer of data between registers

Data is initially entered into the 8-bit register A on the trailing edge of the clock pulse applied to the clock line of the register. The outputs of register A are connected directly to the data input lines of the eight flip-flops in register B. On receipt of a TRANSFER pulse on the clock line of register B, the data held by the A-register is transferred to the B-register. It should be noticed that although a transfer has taken place the contents of register A are unchanged and will remain so until they are cleared, or, alternatively, overwritten by a further data entry.

Such a transfer of data from one register to another can be described by means of the *Register Transfer Language* statement

T: $(B) \leftarrow (A)$

This statement is interpreted as 'The contents of register A are transferred to register B on receipt of the control signal T.'

Serial transfer of data between registers is also possible. Such a transfer is achieved by using the shift mode of operation. Register connections for this mode of operation are illustrated in Figure 2.18. The register consists of eight JK flip-flops, the least significant one having been converted to a D-type flip-flop by the inclusion of an inverter between its J and K lines.

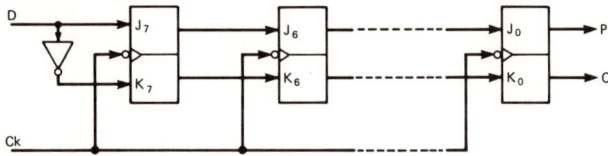

Figure 2.18 A serially operated shift register

It will be assumed that an isolated 1 is to be entered into the most significant flip-flop in the register. On the receipt of the first clock pulse, the 1 is entered into the least significant flip-flop at the left-hand end of the register, and on receipt of the next clock pulse

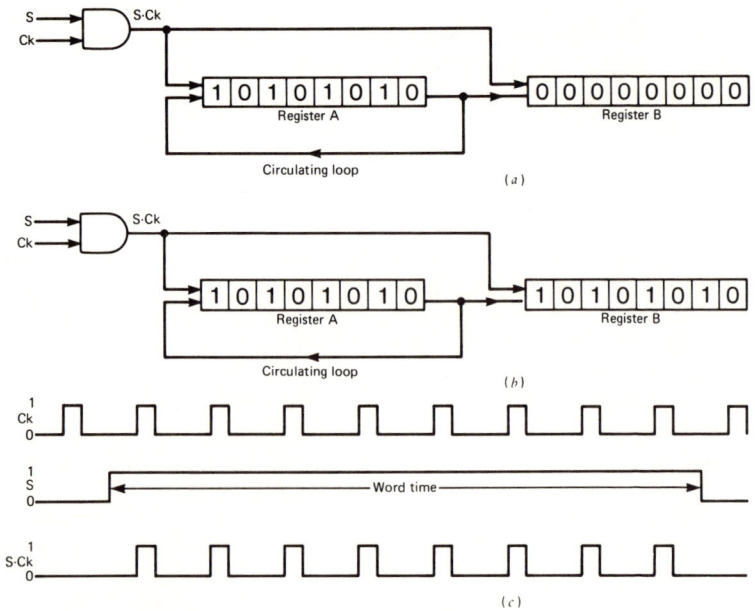

Figure 2.19 Serial transfer of data from register A to register B (a) Before transfer (b) After transfer (c) Timing diagram

it is transferred from the least significant flip-flop to the next one on its right. In all, it will require eight clock pulses to complete the transfer of the 1 to the most significant flip-flop.

If it is required to take the 1 out of the register this can be done by the application of a further clock pulse. In the case described above, the 1 appearing on line P and its inverse appearing on line Q, after the eighth clock pulse, can be transferred to another register or device on the ninth clock pulse. The register would then be operating in the serial in/serial out mode, and such a register would be called a SISO.

Block diagrams describing serial transfer of data from one 8-bit register A to a second 8-bit register B are shown in Figure 2.19. Figure 2.19(a) shows the state of the registers before transfer, and Figure 2.19(b) the state of the registers after transfer. It should be noticed that the data originally in register A is retained by that register via the circulating connection. The transfer of the 8-bit word from register A to register B requires eight clock pulses, and the time duration of these pulses is defined as the *word time*, as illustrated in Figure 2.19(c).

A register transfer language statement that describes the transfer is

S: (B)←(A), (A)←(A)

This statement can be interpreted as meaning that data is transferred from register A to register B as well as being recirculated to A, providing the control signal S has a value of 1. Since S = 1 for a period of eight clock pulses the statement defines the transfer of an 8-bit word.

2.14 The data latching register

A common type of IC register employed in microcomputer systems is the 8-bit data latching register, which employs positive level triggered D-type flip-flops as shown in Figure 2.20(a). Connections to the register are shown in Figure 2.20(b). Data arriving at the D-inputs of the register D_0 to D_7 is transferred to the corresponding outputs of the register Q_0 to Q_7, providing the clock signal is high. On the trailing edge of the clock pulse the data is latched at the Q outputs and remains there until the clock signal goes high again. The asynchronous active low clear signal resets all the flip-flops in the register simultaneously to zero.

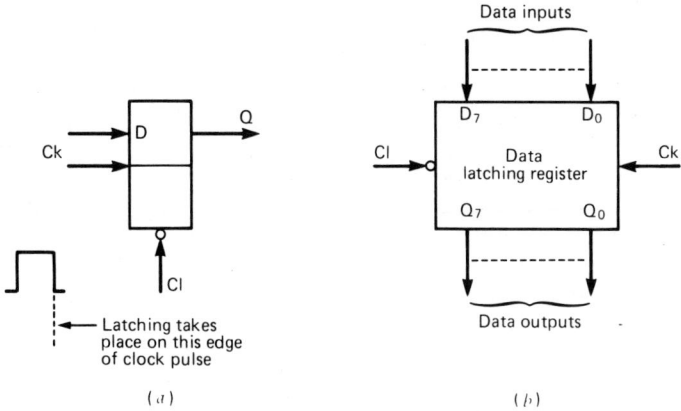

Figure 2.20 (a) Positive level triggered D-type latch (b) 8-bit data latching register

2.15 The tri-state register

Another common type of IC register encountered in microcomputer systems is the 8-bit tri-state register. In this register each individual D-type latch in the register has a tri-state output, as illustrated in Figure 2.21(a). A block diagram for the register is shown in Figure 2.21(b). The outputs 0_0-0_7 operate in the same way as those in the latching register, providing the tri-state enable signal E is high. However, when this signal is low, the outputs of the tri-state gates present a high output impedance. Resetting the register is achieved by an asynchronous active low clear signal.

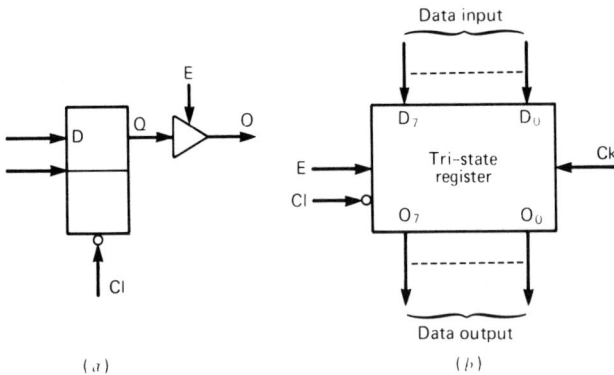

Figure 2.21 (a) Positive level triggered tri-state D-type latch (b) 8-bit tri-state register

2.16 The edge-triggered register

Edge-triggered 8-bit registers are also available in integrated circuit form. They utilize edge-triggered D-type flip-flops, and the data inputs are transferred to the corresponding register outputs on either the leading edge or trailing edge of a clock pulse, depending upon the type of triggering that is in use.

2.17 Data selectors or multiplexers

A multiplexer selects 1-out-of-n data lines as its output where n is typically 4, 8, or 16. Selection of individual data lines is made by means of control signals. For a four-input multiplexer, two control signals are required, while for an eight-input multiplexer three control signals are used for selection, and so on. A block diagram of a multiplexer having four input data lines, D_0, D_1, D_2 and D_3 and two output lines, f and $\bar{f}$ is shown in Figure 2.22(a). The device has two control lines, A and B, and an active low enable line E. When functioning as a data selector, the multiplexer may be regarded as a rotating single-pole switch which selects 1-out-of-4 input lines, as shown in Figure 2.22(b).

Implementation of the multiplexer at gate level is shown in Figure 2.22(c). In essence, the circuit is an AND-OR-INVERT gate having complementary outputs. The Boolean function which represents the output f of the circuit is

$$f = \bar{E} \, (\bar{A}\bar{B}D_0 + \bar{A}BD_1 + A\bar{B}D_2 + ABD_3)$$

Data lines are selected by applying the appropriate binary coded signal to the control lines A and B. When the control signals are A = 0 and B = 0, $\bar{A}\bar{B}$ = 1 and data line D_0 is selected, and when they are A = 0 and B = 1, $\bar{A}B$ = 1 and data line D_1 is selected, and so on. Additionally, if E = 1, the operation of the multiplexer is inhibited.

Some integrated circuit chips contain two or more multiplexers. In the example illustrated in Figure 2.23(a) there are four 2-to-1 multiplexers. Each multiplexer consists of two AND and one OR gate associated with a pair of data lines such as D_{A0} and D_{B0}, as shown in Figure 2.23(b). Since there are only two data lines for each multiplexer, only one control signal X is required to distinguish between them.

An alternative way of looking at a multiplexer is to regard it as a device which converts information presented in a parallel format

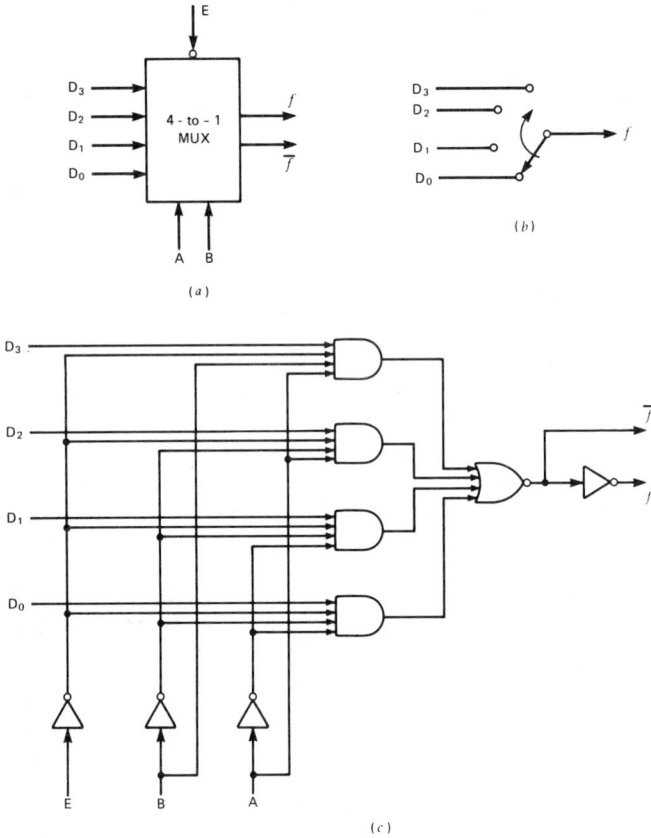

Figure 2.22 (a) Block diagram of a four-input multiplexer (b) Selection of 1-out-of-4 lines by means of a rotating single-pole switch (c) Implementation of a four-input multiplexer

to serial form. Most digital systems process data in parallel form, but if the data has to be transmitted over any distance it is convenient to convert it to serial form so that only a single line is required for its transmission.

2.18 Demultiplexers and decoders

A demultiplexer performs the opposite function to that of a multiplexer in that it can be used to convert a serially transmitted stream of binary digits into parallel form at the end of a

(a)

(b)

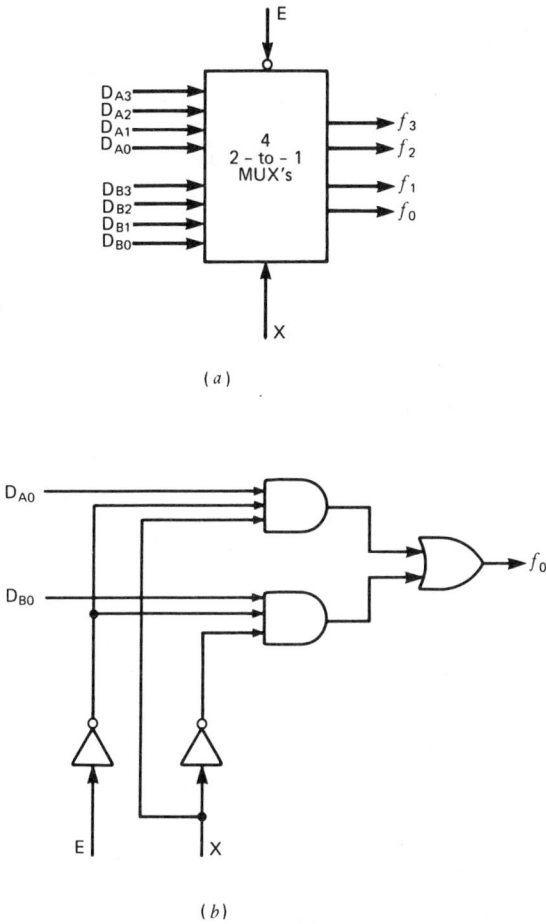

Figure 2.23 (a) Block diagram of chip containing four 2-to-1 multiplexers (b) Logic diagram for one channel of the MUX chip

transmission line. The data is then in a suitable form for transfer to a digital machine for processing. A block diagram of a demultiplexer is shown in Figure 2.24(a). A single data input line D can be connected to any one of four output lines by the appropriate choice of signals on the control lines A and B, providing the enable signal E is active low.

With two control lines there are four possible combinations of A and B, namely 00, 01, 10 and 11, and hence the maximum number of output lines that can be selected is four, as illustrated in the logic diagram of Figure 2.24(*b*).

If $E = 0$ and $D = 0$, then $K = 1$, and for control signals $A = 1$ and $B = 0$ the output $O_2 = A\overline{B}K = 0$. Alternatively, if $E = 0$ and $D = 1$, then $K = 0$, and for control signals $A = 1$ and $B = 1$, the output $O_3 = \overline{AB}K = 1$. In both instances the data is transferred to

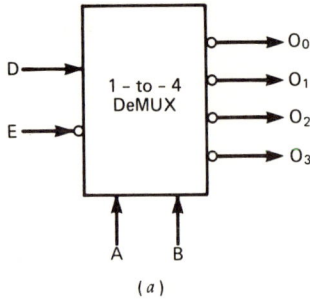

(*a*)

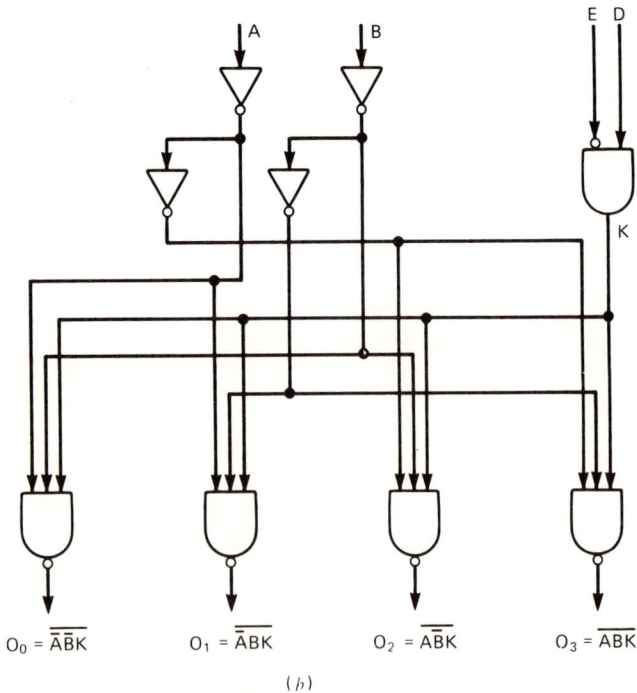

$O_0 = \overline{A}\overline{B}K \qquad O_1 = \overline{A}BK \qquad O_2 = A\overline{B}K \qquad O_3 = \overline{AB}K$

(*b*)

Figure 2.24 (*a*) 1-to-4 line demultiplexer (*b*) Logic diagram the selected output gate by the address applied to the control lines A and B.

A decoder is, in effect, a demultiplexer without a data input. It converts an n-bit code into a maximum of $p = 2^n$ individual outputs. A block diagram for a 3-to-8 line decoder is shown in Figure 2.25(a), while the truth table for the device is shown in Figure 2.25(b). The outputs for this decoder are active low, as indicated by the negation circles, so that if the input code is A = 0, B = 1 and C = 1 the output O_3 will be low, while all other outputs are high.

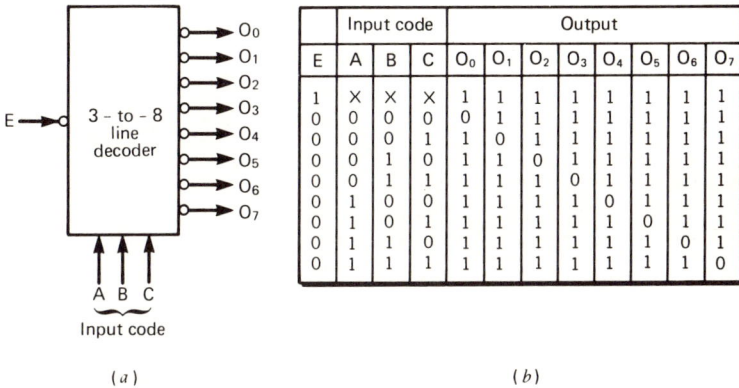

	Input code			Output							
E	A	B	C	O_0	O_1	O_2	O_3	O_4	O_5	O_6	O_7
1	X	X	X	1	1	1	1	1	1	1	1
0	0	0	0	0	1	1	1	1	1	1	1
0	0	0	1	1	0	1	1	1	1	1	1
0	0	1	0	1	1	0	1	1	1	1	1
0	0	1	1	1	1	1	0	1	1	1	1
0	1	0	0	1	1	1	1	0	1	1	1
0	1	0	1	1	1	1	1	1	0	1	1
0	1	1	0	1	1	1	1	1	1	0	1
0	1	1	1	1	1	1	1	1	1	1	0

(a)   (b)

Figure 2.25 (a) 3-to-8 line decoder (b) Truth table

Decoders are widely used in microprocessor systems. A very common requirement in such a system is the selection of one of a number of memory chips by the microprocessor. The 3-to-8 line decoder shown in Figure 2.25(a) may be used to select any one of eight memory chips in accordance with the input code which would be supplied by the microprocessor.

The decoder has a single enable line. However, quite frequently a chip of this kind may have multiple enables, with some of them being active high while others are active low. Control of a variety of devices connected to a bus system has become increasingly more flexible with the introduction of multiple enables.

2.19 Encoders

An encoder performs the opposite function to that of a decoder. It has a number of inputs, and it generates an output code corresponding to the input that has been activated. A block diagram of a four-input encoder is shown in Figure 2.26(a). The device has four inputs, I_0, I_1, I_2 and I_3, and two outputs, A and B. There are four possible combinations of the output signals and each one of these combinations is used to identify a particular input signal when it has been activated.

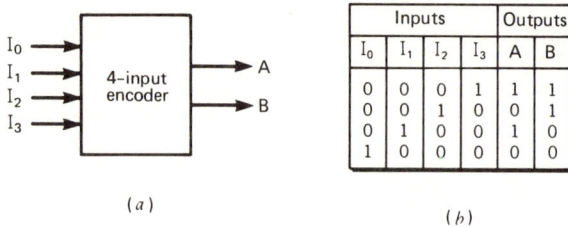

(a)

Inputs				Outputs	
I_0	I_1	I_2	I_3	A	B
0	0	0	1	1	1
0	0	1	0	0	1
0	1	0	0	1	0
1	0	0	0	0	0

(b)

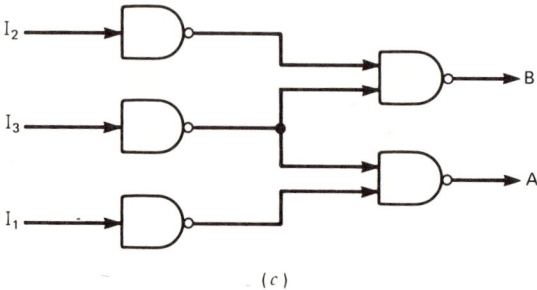

(c)

Figure 2.26 (a) The four-input encoder (b) Truth table (c) Implementation

The truth table for the encoder is shown in Figure 2.26(b), and the following Boolean equations for the outputs A and B have been extracted from this table:

$$A = I_3 + I_1 = \overline{\overline{I_3}\,\overline{I_1}}$$

and

$$B = I_3 + I_2 = \overline{\overline{I_3}\,\overline{I_2}}$$

Implementation of these equations is illustrated in Figure 2.26(c).

The encoder gives a low output on both the A and B lines when all the inputs are low. An extra output line is required to distinguish between this case and the one where $I_0 = 1$ and all other inputs are low.

A modification to the simple encoder just described would allow it to deal with a situation where more than one of the inputs are simultaneously high. Such a modification might be introduced to ensure that, when this situation occurs, the input signal I having the largest subscript has the highest priority. When modified in this way the simple encoder is called a *priority encoder*.

The truth table for the priority encoder shown in Figure 2.27(*a*) is identical to that of the simple encoder except X's are entered in certain cases to identify 'don't care' conditions. For example, the first row indicates that if $I_3 = 1$, $A = 1$ and $B = 1$, irrespective of whether I_0, I_1 and I_2 are present or absent.

Inputs				Outputs	
I_0	I_1	I_2	I_3	A	B
X	X	X	1	1	1
X	X	1	0	0	1
X	1	0	0	1	0
1	0	0	0	0	0

(*a*)

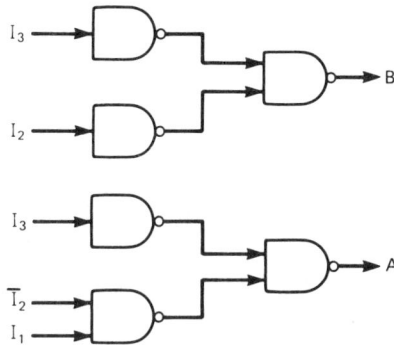

Figure 2.27 (*a*) Truth table for a four-input priority encoder (*b*) Implementation

From the truth table the following Boolean equations are obtained:

$$A = I_3 + \bar{I}_2 \, I_1$$

and

$$B = I_3 + I_2$$

The implementation of these equations is illustrated in Figure 2.27(*b*).

Priority encoders such as the one described above are frequently to be found in microprocessor interrupt systems.

2.20 Parity generation and checking

When digital data is transmitted from one location to another, it is desirable to know at the receiving end whether the received data is error-free. A simple form of error detection can be achieved by adding an extra bit to the transmitted word. The additional transmitted bit is called a *parity bit*.

There are two different parity systems presently in use, even and odd. In an even parity system the parity bit added to the word to be transmitted is chosen so that the number of 1s in the modified word is even. This is illustrated in the following example, where the word $(AE)_{16}$ would require the addition of a 1 to give even parity in the modified word:

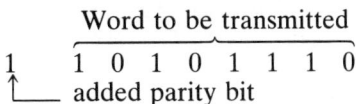

Word to be transmitted

1   1 0 1 0 1 1 1 0
⤹— added parity bit

Alternatively, in an odd parity system, the added parity bit ensures that the modified word contains an odd number of 1s. For the word $(AE)_{16}$ the added parity bit will be a 0 and this maintains the odd parity in the modified word:

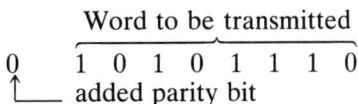

Word to be transmitted

0   1 0 1 0 1 1 1 0
⤹— added parity bit

The truth table for a 3-bit even/odd parity generator is shown in Figure 2.28(a), where D_2, D_1 and D_0 represent the data bits to be transmitted, and p_0 and p_e represent the odd and even parity bits to be generated by the parity generation circuit.

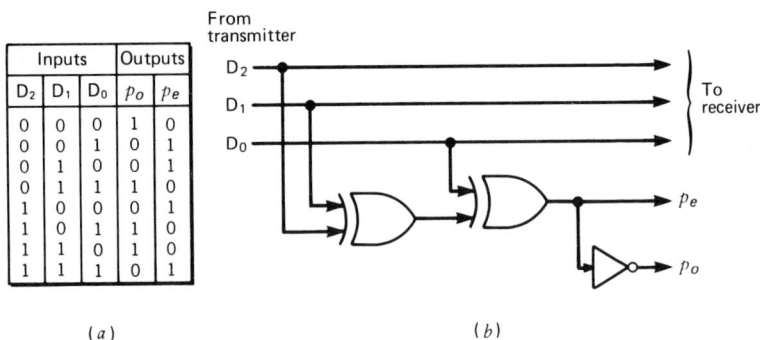

Inputs			Outputs	
D_2	D_1	D_0	p_o	p_e
0	0	0	1	0
0	0	1	0	1
0	1	0	0	1
0	1	1	1	0
1	0	0	0	1
1	0	1	1	0
1	1	0	1	0
1	1	1	0	1

(a)   (b)

Figure 2.28 (a) Truth table for parity generator (b) Implementation

The Boolean equation for p_e, extracted from the truth table, is

$$p_e = \bar{D}_2\bar{D}_1D_0 + \bar{D}_2D_1\bar{D}_0 + D_2\bar{D}_1\bar{D}_0 + D_2D_1D_0$$

and this can be manipulated algebraically to give

$$p_e = D_2 \oplus D_1 \oplus D_0$$

An examination of the truth table shows that p_o is the inverse of p_e, so that

$$p_o = \overline{D_2 \oplus D_1 \oplus D_0}$$

The implementation of these equations is shown in Figure 2.28(b).

The addition of extra data bits merely adds additional Exclusive-OR terms to the p_e and p_o equations. If the word to be transmitted contains four digits rather than three then the parity generation equations would become

$$p_e = D_3 \oplus D_2 \oplus D_1 \oplus D_0$$

and

$$p_o = \overline{D_3 \oplus D_2 \oplus D_1 \oplus D_0}$$

When the transmitted bits arrive at the receiving end, a circuit called a *parity checker* is used to check the parity of the modified word. In an even parity checking circuit the output is high when there is an error in the parity; similarly, the output of an odd parity checking circuit is high when a parity error is detected.

A truth table for the two types of parity checking circuit is given in Figure 2.29(a) and the parity checking function for an even parity system F_e, extracted from the truth table, is given by the Boolean equation

$$F_e = D_2 \oplus D_1 \oplus D_0 \oplus p$$

while the parity checking function for an odd parity system F_o, by inspection of the truth table, is clearly the inverse of F_e. Hence

$$F_o = \overline{D_2 \oplus D_1 \oplus D_0 \oplus p}$$

The implementation of the parity checking circuit is illustrated in Figure 2.29(b).

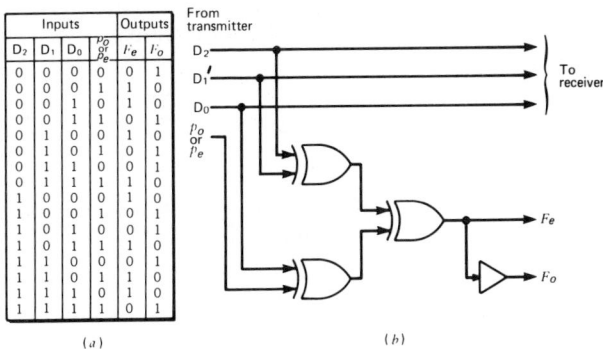

Inputs				Outputs	
D_2	D_1	D_0	p_o or p_e	F_e	F_o
0	0	0	0	0	1
0	0	0	1	1	0
0	0	1	0	1	0
0	0	1	1	0	1
0	1	0	0	1	0
0	1	0	1	0	1
0	1	1	0	0	1
0	1	1	1	1	0
1	0	0	0	1	0
1	0	0	1	0	1
1	0	1	0	0	1
1	0	1	1	1	0
1	1	0	0	0	1
1	1	0	1	1	0
1	1	1	0	1	0
1	1	1	1	0	1

(a)   (b)

Figure 2.29 (a) Truth table for parity checking circuit (b) Implementation

As a general rule in a digital system where the transmission distance is relatively short, it may be assumed that the probability of a single-bit error is small, and that of a double-bit error and higher order errors is extremely small. The parity error detection system just described will detect any odd number of errors, but it cannot detect any even number of errors because such errors will not disturb the parity of the transmitted group of bits. When the probability of double-bit and higher order errors is very small this type of error detection system can serve a useful purpose.

2.21 Comparators

The basic logical comparison element is the Exclusive-NOR gate. Earlier in this chapter it was shown, in Figure 2.1(a), that the output of the Exclusive-NOR gate is high providing both inputs are low, or alternatively both inputs are high. This is indicated in Figure 2.30 where E = 1 if A = 0 and B = 0, or alternatively if A = 1 and B = 1. However, most comparators are required to

$$E = \overline{A}\overline{B} + AB$$

Figure 2.30 The Exclusive-NOR gate

indicate more than equality. They usually have three outputs, one for A > B, a second for A = B, and the third for A < B. A suitable implementation for a single-bit comparator not using an Exclusive-NOR gate and which gives an output for all three conditions is illustrated in Figure 2.31.

In practice, the usual problem is the comparison of two multi-digit words such as A = $A_3A_2A_1$ and B = $B_3B_2B_1$. To compare two such words it is necessary to develop an algorithm which can then be used as the basis of a hardware implementation.

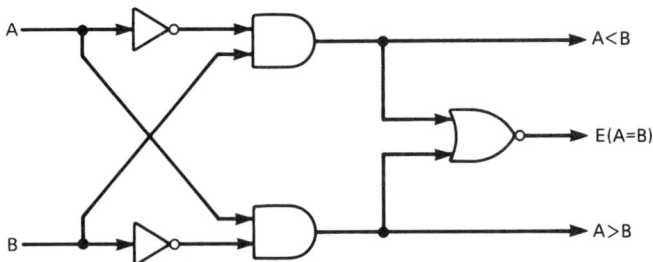

Figure 2.31 A single-bit comparator

2.22 Arithmetic circuits

When adding together two binary digits A and B and a carry C_{in} from a previous stage of the addition, a full adder is required. The block diagram of the full adder circuit is shown in Figure 2.32(a) and it has three inputs. These are the two input digits A and B and the carry-in C_{in}. Additionally, the circuit has two outputs, the sum S and the carry-out C_o to the next most significant stage of the addition. In the least significant stage of any addition $C_{in} = 0$ and in such a case the C_{in} terminal would be connected to ground.

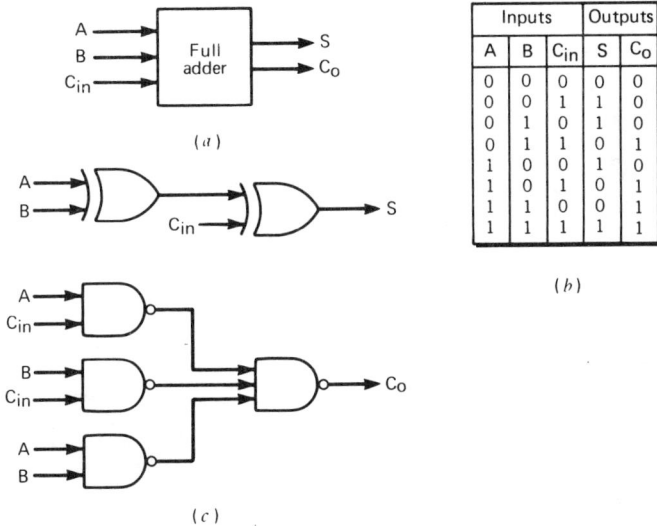

(a)

Inputs			Outputs	
A	B	C_{in}	S	C_o
0	0	0	0	0
0	0	1	1	0
0	1	0	1	0
0	1	1	0	1
1	0	0	1	0
1	0	1	0	1
1	1	0	0	1
1	1	1	1	1

(b)

(c)

Figure 2.32 (a) Block diagram of a full adder (b) Truth table (c) Implementation

The truth table for the full adder is given in Figure 2.32(b). There are eight rows in the table, each of which corresponds to one of the eight possible combinations of the input signals. For each combination the binary sum is found. The sum may be a single-digit number or a two-digit number. In the latter case a carry has been generated and this is entered in the carry-out column.

From the table the following Boolean equations for the sum S and the carry out C_0 are obtained:

$$S = \bar{A}\bar{B}C_{in} + \bar{A}B\bar{C}_{in} + A\bar{B}\bar{C}_{in} + ABC_{in}$$

and

$$C_o = \bar{A}BC_{in} + A\bar{B}C_{in} + AB\bar{C}_{in} + ABC_{in}$$

These equations can be reduced to

$$S = A \oplus B \oplus C_{in}$$

and

$$C_o = AC_{in} + BC_{in} + AB$$

The implementation of the full adder using these equations is illustrated in Figure 2.32(c).

It is clearly now a simple matter to build, for example, a 4-bit parallel adder using the single-bit adder just described as the basic component. A block schematic for such a multi-bit adder is illustrated in Figure 2.33(a). The four separate full adders in this diagram can be combined to form a 4-bit MSI adder chip, as shown in Figure 2.33(b). An adder module of this type will normally constitute a part of the arithmetic/logic unit (ALU) which will be found on all microprocessor chips.

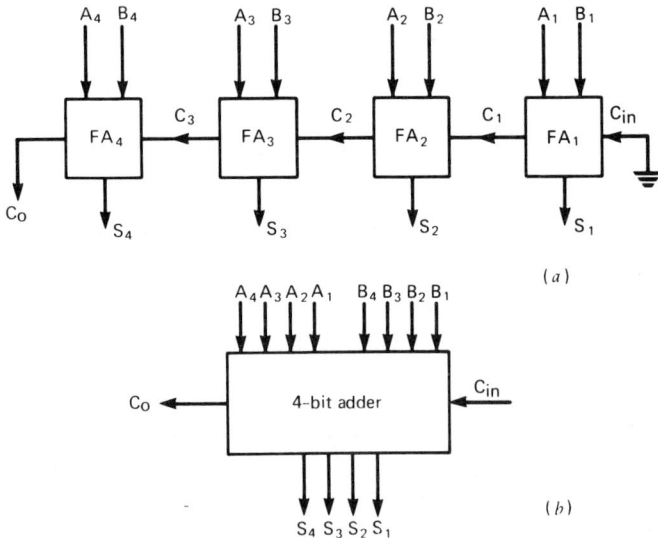

Figure 2.33 (a) A 4-bit parallel adder (b) A 4-bit MSI adder

A 4-bit adder, such as the one shown in Figure 2.33(b) is very versatile, and this unit can perform a number of different arithmetic operations by controlling one set of its inputs. This control can be achieved by inserting a True/Complement unit between the B input lines and the adder. Besides the true/ complement facility, the unit may also provide an output that consists of all the 0s or all the 1s.

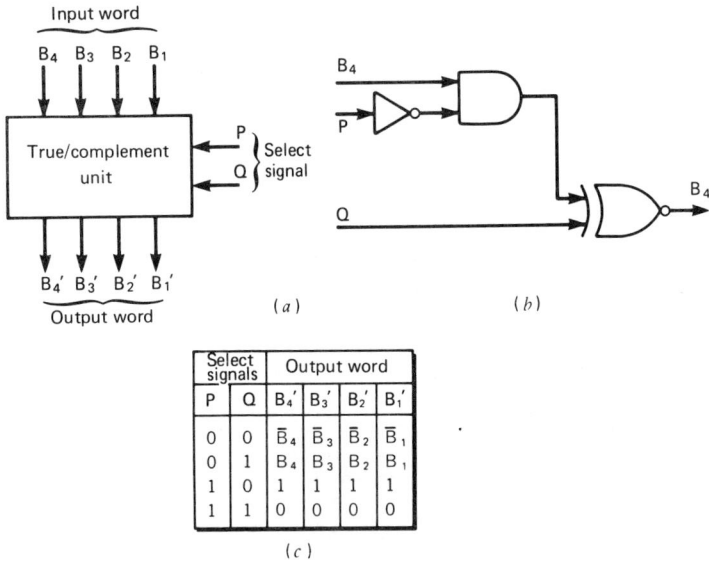

Figure 2.34 (a) Block diagram of True/Complement unit (b) Logic diagram for one channel of the unit (c) Truth table

A block diagram of the True/Complement unit is shown in Figure 2.34(a) and the logic circuit for one of the channels of this unit is shown in Figure 2.34(b). The functional behaviour of the circuit is controlled by the two select signals P and Q. When P = Q = 0 the input word B is inverted and transferred to the output, and when P = 0 and Q = 1 the input word B is simply transmitted to the output. For the other combinations of the select signals, P = 1 and P = 0, the output from each channel is a 1, and for P = Q = 1 the output from each channel will be a 0. The truth table summarizing this circuit behaviour is shown in Figure 2.34(c).

A block diagram of a 4-bit adder operating in conjunction with the True/Complement unit to form a controlled adder is shown in Figure 2.35(a). The functional behaviour of this circuit depends upon the logical value of the two select signals P and Q and the presence or absence of the carry input. There are eight possible combinations of these three input signals and the behaviour of the controlled adder for each of these combinations is illustrated in the eight separate block diagrams shown in Figure 2.35(b). The functional outputs of the controlled adder for each of the eight combinations of P, Q and C_i are tabulated in the function table shown in Figure 2.35(c).

58   Logic

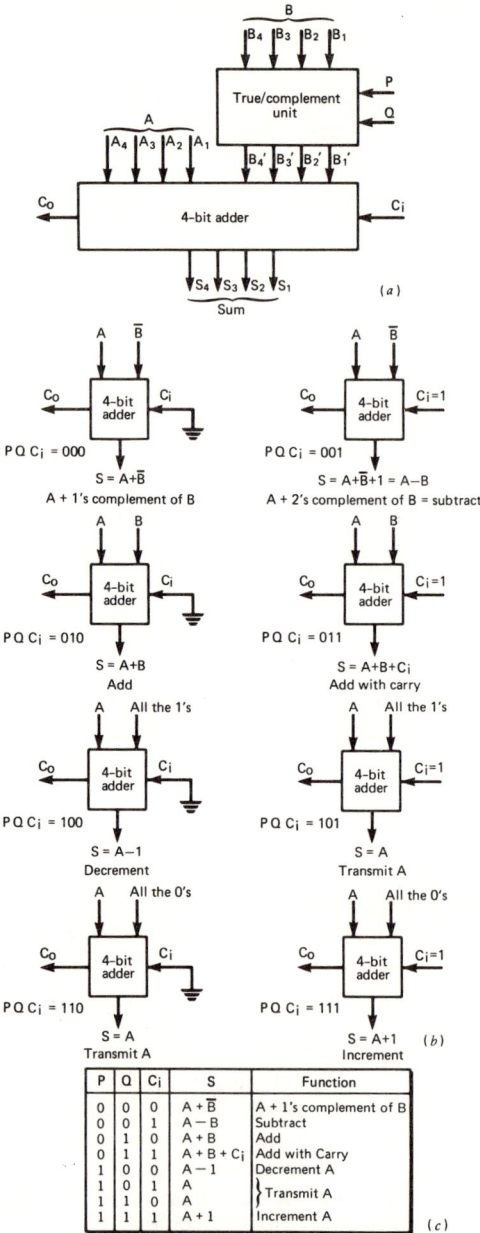

Figure 2.35 (*a*) Block diagram of a controlled adder (*b*) Functional operation (*c*) The function table

2.23 Arithmetic/logic unit

Virtually all microprocessors have an ALU in which both the arithmetic and the logic operations are performed. The minimum hardware requirements of an ALU are a controlled adder operating in conjunction with a pair of registers and also the logic circuitry necessary to implement basic logic functions such as AND and OR, etc.

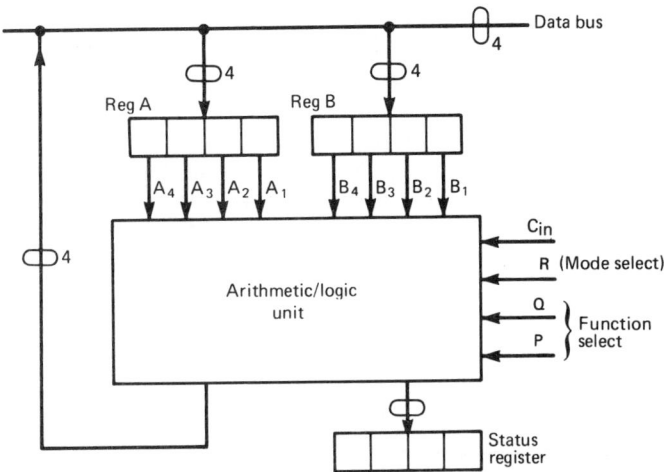

Figure 2.36 Block diagram of arithmetic/logic unit

A block diagram of a 4-bit ALU is shown in Figure 2.36. The two registers A and B are connected directly to a common bus system and can be loaded separately with data from this system. The mode select signal R decides whether the ALU is going to perform an arithmetic or a logic operation on the data, while the select signals P and Q select the appropriate logic or arithmetic function to be implemented. After the arithmetic or logic operation has been performed on the data in the ALU the result is returned via the data bus to the A register, thus allowing further arithmetic or logic operations to be performed on the result.

The ALU will also generate status bits which are in turn transferred to a status register. The status bits provide information regarding the result of the operation just performed in the ALU. For example, if the operation had been an addition and a carry-out had been generated, then it would be regarded as a status bit and would be stored in the status register. Alternatively, if the result of the operation had been zero, then a zero status flag would be generated and transferred to the status register. The ALU in Figure 2.36 has four status bits, but it should be noted that the number of status bits varies from machine to machine, as does the information they provide. Status bits are very important in microprocessor operation, since they can usually be examined and a program jump can be initiated on the basis of that examination.

The arithmetic section of the ALU is, in effect, a controlled adder and has been dealt with in some detail previously. However, the logic section of the device consists of groups of AND, OR, XOR and NOT gates. Since the ALU illustrated in Figure 2.36 has a pair of 4-bit inputs it will contain groups of four of each of the gates specified earlier in this paragraph. Implementation of one channel of the logic section of the ALU is shown in Figure 2.37(a). The required logic function is selected by the multiplexer control signals which can be regarded as function selection signals. A function table for the logic section of the ALU is shown in Figure 2.37(b).

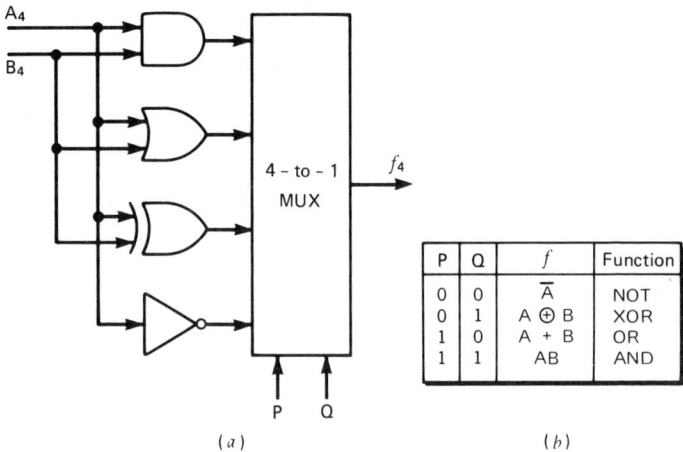

P	Q	f	Function
0	0	$\overline{A}$	NOT
0	1	$A \oplus B$	XOR
1	0	$A + B$	OR
1	1	AB	AND

Figure 2.37 (a) A single channel of the logic section of the ALU (b) The function table for the logic section

The logic section and arithmetic sections of the ALU are selected by the mode select signal R. This signal can be used as the control signal for a two-input multiplexer. Selection of one channel of either the arithmetic or logic sections of the ALU by

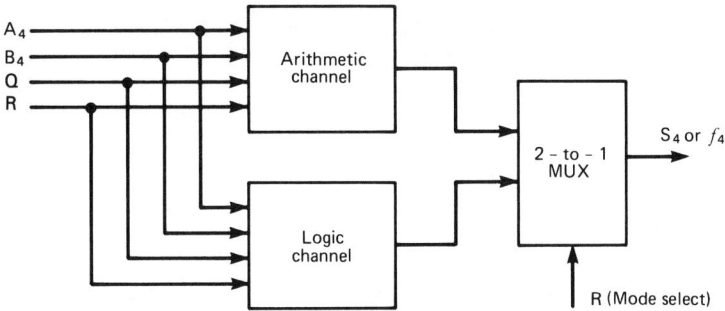

Figure 2.38 Block diagram of a single channel of the ALU

R	P	Q	C_i	f or S	Function
0	0	0	X	$f = \bar{A}$	NOT
0	0	1	X	$f = A \oplus B$	XOR
0	1	0	X	$f = A + B$	OR
0	1	1	X	$f = AB$	AND
1	0	0	0	$S = A + \bar{B}$	A + 1's complement of B
1	0	0	1	$S = A - B$	Subtract
1	0	1	0	$S = A + B$	Add
1	0	1	1	$S = A + B + C_i$	Add with Carry
1	1	0	0	$S = A - 1$	Decrement A
1	1	0	1	$S = A$	} Transmit A
1	1	1	0	$S = A$	
1	1	1	1	$S = A + 1$	Increment A

Figure 2.39 Function table for the ALU

the mode select multiplexer is illustrated in Figure 2.38. The output from the channel in this diagram will be S_4 if the arithmetic channel is selected, and f_4 if the logic channel is selected.

A function table for the ALU, incorporating both logic and arithmetic functions, is given in Figure 2.39.

Problems

2.1   Determine the output equation of the gate circuits in Figures 2.40(a) and (b).

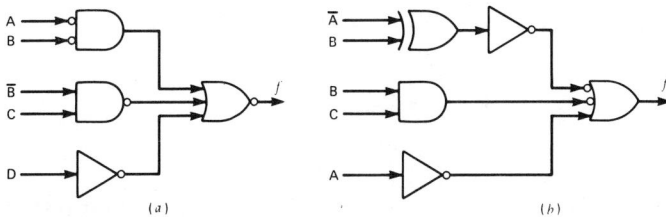

Figure 2.40 (a) Gate circuit 1 (b) Gate circuit 2

2.2   A D-type flip-flop is defined as one in which the input signal
on the D-line is transferred to the output on the trailing edge
of the first complete clock pulse after the change in the input
signal. Design a D-type flip-flop that satisfies this specification using asynchronous circuit techniques and implement
the design using NAND gates.

2.3

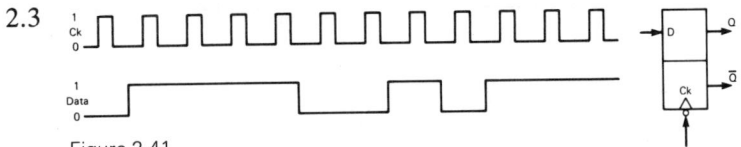

Figure 2.41

Determine the output waveform of the D-type flip-flop
shown in Figure 2.41 and compare it with the input data.
How can an additional delay of one clock period be
achieved?

2.4   Develop a circuit arrangement consisting of two registers X
and Y which allows the parallel transfer of data from register
X to register Y. Each register is four bits long.

2.5   Draw a block diagram to implement the following register
transfer language statements

$$A \leftarrow A + B \qquad O \leftarrow C_n \oplus C_{n+1}$$

where A and B are both 4-bit registers and O is a single-bit
register.

2.6   A block diagram for a circuit used to add three binary digits
P, Q and R is shown in Figure 2.42. Determine the Boolean
equations for S, Co' and Co'' and implement them using

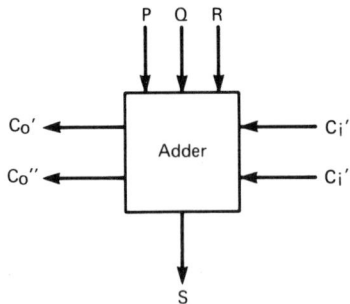

Figure 2.42

(a) A 16-to-1 multiplexer, and
(b) an 8-to-1 multiplexer and NAND-implemented residue
logic.

2.7   A register having a parallel output contains a 16-bit word. The word has to be converted into serial form using a group of 4-to-1 multiplexers. Draw a block diagram of a suitable system.

2.8   Draw a logic diagram for a 2-to-4 line demultiplexer, with an enable input, using NOR gates. Show how a system of 2-to-4 line demultiplexers can be used to convert a 16-bit serially transmitted word into parallel form.

2.9   A microprocessor services eight different peripherals. A peripheral indicates to the processor that it requires service by raising a flag. Develop a combinational encoder circuit that will identify the peripheral requiring service. Assume that priority is defined by the subscript attached to the flag and that the flag with the highest subscript has highest priority.

2.10 A half-adder is designed to add together two binary digits A and B and generate two outputs S (Sum) and C (Carry). Assuming that only uncomplemented variables are available, design the half-adder using no more than five NAND gates or five NOR gates, but not both.

   Using two half-adders and two NOR gates implement a full adder circuit.

2.11 Design an adder/subtractor circuit having two 4-bit binary inputs A and B. When the mode control $M = 0$ the circuit performs the addition $A + B$ and when $M = 1$ the circuit performs the subtraction $A - B$ using 2's complement arithmetic.

2.12 Design an arithmetic circuit which is to be controlled by two mode variables M_0 and M_1. The arithmetic operations to be performed by the circuit are given by the table below for specified values of M_0 and M_1.

M_0	M_1	$C_{in} = 0$	$C_{in} = 1$
0	0	$O = A + B$	$O = A + B + 1$
0	1	$O = A$	$O = A + 1$
1	0	$O = \bar{B}$	$O = \bar{B} + 1$
1	1	$O = A + \bar{B}$	$O = A + \bar{B} + 1$

3 Memory

3.1 Introduction

The operation of all computing systems depends upon the stored program concept, and consequently memory is an important feature of the system. Computation time is very dependent upon memory retrieval, and storage times and system costs are largely determined by memory capacity. It is not surprising therefore that memory has been the subject of much development work in recent years.

Memory in the most general sense can be divided into two distinct groups:

(1) internal memory,
(2) external memory.

Internal memory is continuously in communication with the central processing unit and the rest of the computer as it executes a program. It is used for storing data and instructions, and memory access time (the phrase used for storage and retrieval) is fast.

External memory is a mass storage system external to the computer and normally slower than internal memory. It is capable of storing large blocks of data (millions of bits) and is used for data and information prior to transfer to the internal memory of the machine. Additionally, it is capable of receiving information and data from the machine for long-term storage.

Advances in LSI technology have made it possible to get a large number of single-bit memory elements onto one chip, thus giving the chip a large storage capacity. The active devices around which these single-bit memory elements have been developed are the bipolar and MOS transistors. The memory chips are fast, and the cost of production has decreased rapidly in recent years. It is now general practice for microcomputer systems to use bipolar or MOS devices to fulfil their internal memory requirements.

Semiconductor memories, although admirably suited for internal memory, are prohibitively expensive for external mass storage systems. For this purpose magnetic tapes, disks and drums are widely used, although recent years have seen the advent of charge-coupled devices and magnetic bubble memories.

This chapter will deal with those devices which constitute the internal memory of a microcomputing system. Basic memory concepts will be examined, and the various types of memory in use and their properties will be discussed. Additionally, the principles of memory system organization will be considered.

3.2 The register representation of memory

An IC memory chip consists of an array of addressable registers of identical length, each register occupying a location which has a unique address provided by an address decoder. A block schematic of a typical memory chip is illustrated in Figure 3.1. The memory shown in this diagram has n address lines, and since there are 2^n combinations of n binary digits the chip will house 2^n registers, each addressed by one of the 2^n output lines of the decoder.

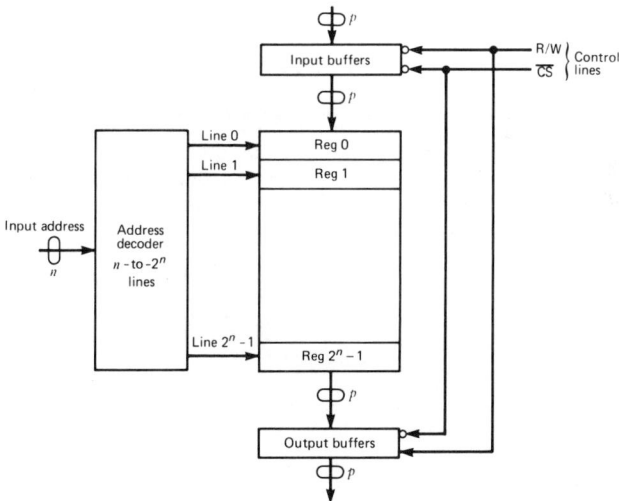

Figure 3.1 The register representation of a memory chip

If each of the registers contains a p-bit word then the capacity of the memory is equal to $p \times 2^n$ bits. For a word length $p = 8$ and a number of address lines $n = 8$ the storage capacity is $8 \times 2^8 = 2048$ bits. Since a group of eight binary digits is commonly referred to as a byte, the storage capacity of this memory is 256 bytes. Increasing the number of address lines will increase the storage capacity; for example, if $n = 10$ then the storage capacity is 1024 bytes and the memory is referred to as a 1 K byte memory, where $1 K = 1024$, $2 K = 2048$, and so on.

For an 8-bit microprocessor system the word length is 8 bits while the address length is 16 bits. Since both word length and addresses are multiples of four bits it is common practice to represent memory location contents and addresses in hexadecimal notation.

The use of the hexadecimal system in this context has a number of advantages. It requires much less clerical work when tabulating memory contents and addresses. It is also difficult to detect errors in strings of binary digits. The more compact hexadecimal representation with its larger range of digits is easier to scan for errors.

The registers within the memory are subject to two operations, reading and writing. The read operation consists of addressing the required location, reading its contents, and then transferring those contents via the output buffers to the outside world using the system bus. The write operation is the reverse of the read operation in that data from the bus is transferred via the input buffers to the addressed register thus overwriting the data presently stored in that register.

Read and write operations are carried out under the control of the Read/Write and Chip Select signals. The memory illustrated in Figure 3.1 is selected by an active low chip select signal $\overline{CS}$, and then if R/W = 1 the output buffers are enabled and data can be read from the addressed location. Alternatively, if R/W = 0 the input buffers are enabled and data can be read into the addressed location. In practice it would be unusual to have separate input and output buffers; common buffers would be used for both functions, thus saving on pin connections.

3.3 Memory timing

The time between the arrival of the address signal and either the appearance of the data at the output terminals or the completion of a write operation is called the *memory access time*. This depends upon the technology used and the structure of the memory. For a memory of the kind shown in Figure 3.1 the access time for two different memory locations selected at random is essentially the same, and for this reason it is called a random access memory (RAM).

Access time can be divided into two separate sections, as indicated in Figure 3.2. The time between the arrival of the address signal and the location of that address signal is called the *latency*, while the time for reading out of or writing into that address is called the *data transfer time*. After the transfer of data a finite time is required for the device to return to its original internal state, and this is called the *settling time*.

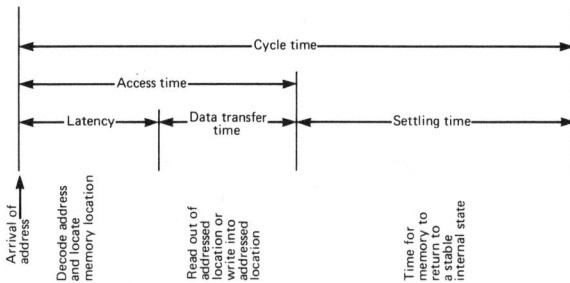

Figure 3.2 Memory timing diagram

The sum of the latency, data transfer time and settling time is called the *cycle time*. This may be defined as the time taken for the memory to complete a read or write operation and then return to its original state. In effect, cycle time is a measure of memory speed.

3.4 Memory classification

Memory can be broadly divided into two classes: (a) read only memory (ROM) and (b) read/write memory (RWM).

Read/write memories are normally referred to as RAM, but is should be recognized that a ROM is also a random access memory and the exclusive usage of this term for read/write memories is not strictly correct.

ROM is used for program storage and invariable data patterns. It is used in situations where memory contents are not required to change and a main area of application for ROM is in dedicated microprocessor systems.

A ROM functions as a memory array whose contents once programmed are permanently fixed and cannot be altered. It is a non-volatile memory, which means that if there is a loss of power or some other circuit malfunction the contents of the ROM are not destroyed. Because the state of a memory cell in a ROM is permanently defined its structure is considerably simpler than that of a memory cell in RWM.

There are four categories of ROMs presently available.

(1) *Mask programmed when manufactured.* The contents are programmed by the manufacturer during fabrication according to customer specification. It is not possible to change the program after packaging.

(2) *PROMs, or programmable ROMs.* The contents are written into a PROM by the user with the aid of a PROM programmer. Once programmed, the contents of the memory are permanently fixed.

(3) *EPROMs, or erasable PROMs.* The contents are programmed by the user but can be subsequently erased by an application of ultraviolet light through a window on the upper surface of the memory chip.

(4) *EAROMs, or electrically alterable ROMs.* This type of ROM can have its program erased by electrical means while still in circuit.

Both EPROMs and EAROMs can be reprogrammed and used again after erasure.

RWMs are used in microcomputer systems for the temporary storage of data and intermediate results of calculations. This type of memory consists of an array of active volatile elements which will lose the stored information if there is a power failure or if power is removed in the normal way.

There are two types of RWMs. Static RWMs store each bit of information in a flip-flop and the information is retained in the memory cell until the power is turned off. In dynamic RWMs the information is stored in the form of an electric charge on a capacitive element. This charge leaks away with time and the element must be periodically refreshed.

The main advantage of dynamic RWMs is their greater packing density. For a given volume, the storage capacity of a dynamic RWM is much greater than for a static RWM. The obvious disadvantage of this device is the requirement for refresh circuits. Since memory manufacturers produce dynamic RWMs with nothing other than 1-bit word lengths, the minimum memory size for an 8-bit microprocessor requires eight separate chips.

3.5 ROMs

Both bipolar and MOS technology are used in the fabrication of ROMs. The significant performance difference between the two technologies is the access time. Bipolar ROMs have access times as low as 50 ns while access times for MOS ROMs are in the range of 500 ns to 1 μs.

The array of registers shown in Figure 3.1 is frequently called the memory matrix. It is conventionally represented by a set of intersecting lines, as illustrated in Figure 3.3. The vertical lines in the diagram are called the word lines and are connected to the outputs of the address decoder, while the horizontal lines called the bit lines are connected to the inputs of the output buffers.

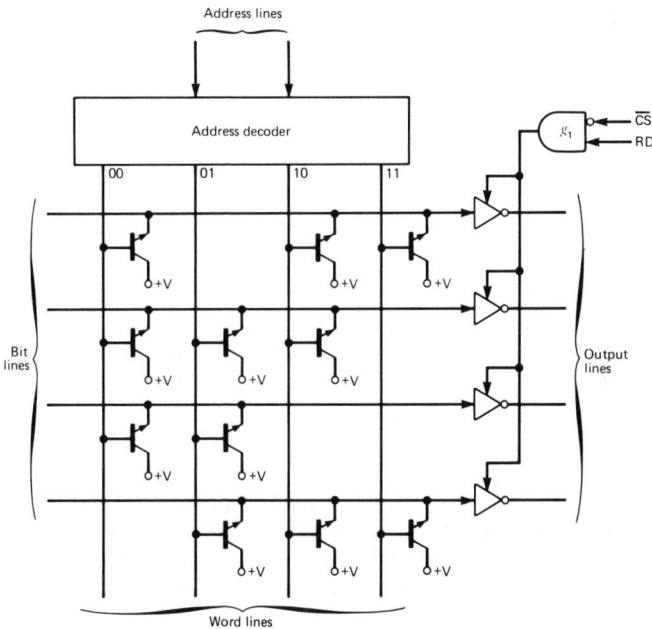

Figure 3.3 A 4 × 4-bit bipolar ROM addressed in one dimension only

Words are programmed into the ROM at each address and the output on a bit line depends on whether it is electrically connected to the addressed word line or not, this connection being made in the example shown via an N-type bipolar transistor. If the connection exists the bit line is raised to a 1 and, if not, it remains at 0. In this way the programmed word at the selected address is transferred to the inputs of the output buffers, which are enabled when the chip select signal is low and the read signal is high.

In the example shown in Figure 3.3, if the address is 10 the 4-bit word transferred to the output is 1101 providing the buffer enable signals generated by gate g_1 are high.

The ROM shown in Figure 3.3 is addressed in one dimension only. However, as the number of address lines is increased, the number of outputs from the address decoder entering the ROM matrix rises dramatically. If there are five address lines the decoder has $2^5 = 32$ outputs entering the ROM matrix, but if the number of address lines is doubled to ten the decoder has $2^{10} = 1024$ outputs entering the matrix. In order to reduce the total number of leads entering and leaving the matrix a two-dimensional addressing scheme can be used.

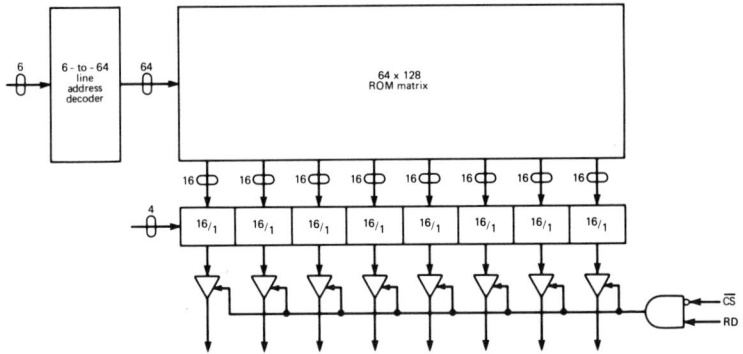

Figure 3.4 A 1 K byte two-dimensionally addressed ROM

A two-dimensionally addressed ROM is shown in Figure 3.4. The most significant six address digits are used to select 1-out-of-64 rows in the memory matrix, while the four least significant address digits are used as control signals for eight 16-to-1 multiplexers connected to the 128 bit lines. The bit lines are in groups of 16 and one of the 16 bit lines fed to each of the multiplexers is selected by the four control signals. The outputs of the multiplexers are then connected to the inputs of the tri-state buffers which are enabled when the logic signal $E = \overline{CS}.RD$ is high.

The total capacity of the memory is $64 \times 128 = 8192$ bits, and since the output words are 8 bits long the memory is more conveniently called a 1 K byte memory. For this memory the number of lines entering and leaving the matrix is $64 + 128 = 192$.

If, however, single-dimensional addressing had been used, the number of lines entering and leaving the matrix would have been 1032.

3.6 PROMs

PROMs are fabricated with bipolar transistors employed as the active element. A fusible link is connected in series with the transistor between the bit and word lines as shown in Figure 3.5. Programming is achieved by passing a sequence of current pulses through the link until it has blown. When the fusible link is intact a 1 is stored, and when it has been blown a 0 is stored.

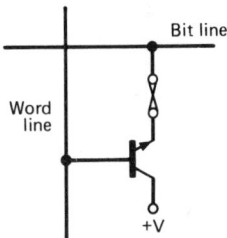

Figure 3.5 Fusing arrangements for bipolar ROM

If the word line associated with the transistor is selected it is turned on by the potential applied to the base. The transistor is heavily conducting, so that the voltage between collector and emitter is approximately zero and the voltage V is transferred to the bit line.

The most commonly used fuse technologies are nichrome or polycrystalline silicon fusing. Unfortunately the fusing technique cannot be used with MOS technology because the current levels required during the fusing operation are incompatible with MOS impedance levels and thus the programming of a MOS device is not readily achieved.

PROMs are more expensive than ROMs and also more complex because of the introduction of the fusible link into the cell structure during fabrication. However, they do not require the preparation of a mask prior to manufacture. They are used in the early stages of product development when considerable changes may have to be made. When the design has been finalized a mask can be prepared for the manufacture of a ROM which, when programmed, will be dedicated to the proven design.

PROMs are programmed with the assistance of a PROM programmer, which is an instrument that generates the address, the data, and the programming pulses required to program the PROM. There are two types of instrument available. The first is a dedicated instrument that is satisfactory for one type of PROM only. Alternatively, there is a more versatile instrument that can be modified for programming a number of different types of PROM by the insertion of an appropriate module. This module adapts the programmer to the specified programming voltages, currents and pin configurations required by the memory chip to be programmed.

3.7 EPROMs

The active device in the memory matrix of an EPROM is a floating gate, avalanche injection, charge storage device. Essentially it is a silicon gate MOSFET with no electrical connection to the gate. The advantage of the EPROM is that data can be stored in the memory matrix for long periods of time, but, if necessary, the contents of the memory can be erased and the device reprogrammed.

To store a 1 in one of the single-bit memory elements in the array a voltage pulse is applied between drain and source of the transistor. This results in the storage of electrons by an avalanche injection process on its floating gate. Presence or absence of stored charge can then be detected by measuring the conductance between source and drain. In effect, the operation of this type of memory relies on the high impedance property of the memory cells which governs the retention of stored charge.

Because the gate is floating the stored charge cannot be removed by electrical methods. However, the application of ultraviolet light to the array initiates a photo-electric current between the gate and silicon substrate which removes the stored charge. A piece of transparent quartz is fitted over the memory array and seals the package. The ultraviolet light is directed onto this window if erasure of the stored data is required.

3.8 RWMs

Read/write memory elements for microcomputer systems are used for storing data or intermediate results. They are either bipolar or MOS semiconductor devices and for the static type of RWM they consist of an array of flip-flops on a single IC chip. In the case of dynamic RWMs they consist of an array of capacitors which can be charged or discharged through an electronically controlled switch.

The single-bit memory elements can be addressed in one or two dimensions. Large RWMs make use of a two-dimensional addressing scheme such as the one illustrated in Figure 3.6. The Y

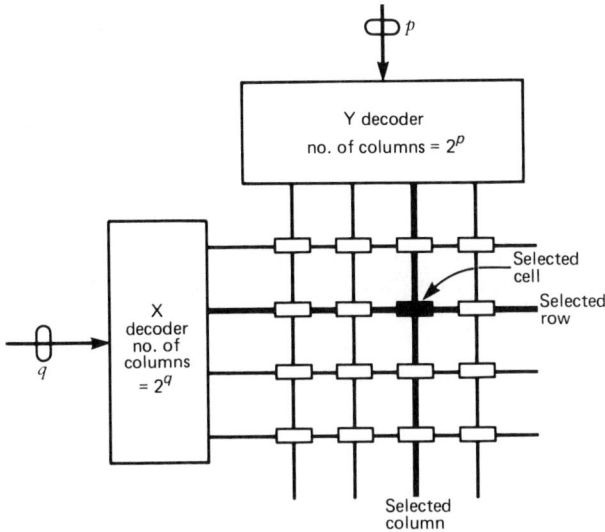

Figure 3.6 Schematic diagram of a bit-organized store

address decoder selects a column in the memory matrix while the X address decoder selects a row, and the intersection of the selected row and column locates the memory cell to be accessed. When a single cell is addressed in this manner the memory is said to be bit organized, but when several cells are accessed simultaneously, as shown in Figure 3.1, the memory is said to be word organized.

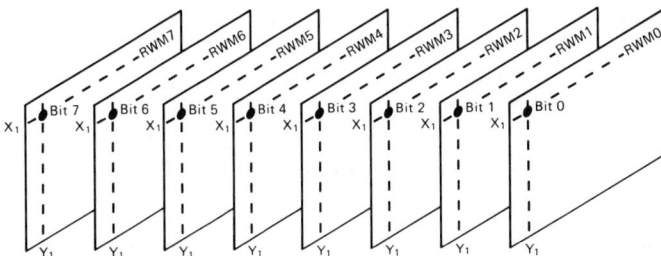

Figure 3.7 Storage of an 8-bit word using eight bit-organized memory planes

To store an 8-bit word when the memory is bit organized requires eight RWM planes, as illustrated in Figure 3.7. Each memory plane contributes one bit to a stored word and these bits are stored at identical locations on the individual memory planes.

3.9 The static MOS RWM

The circuit shown in Figure 3.8 is that of a typical NMOS single-bit memory cell, where transistors T_1 and T_2 are the active devices of a bistable element and transistors T_3 and T_4 are the loads for T_1 and T_2. Transistors T_5 and T_6 are the access transistors and they will allow bidirectional current flow between the complementary bit lines B and $\bar{B}$ and the transistors T_1 and T_2. To access the cell for either reading or writing the potential of the word line W is raised, thus switching on the access transistors T_5 and T_6 and making a direct connection between the points D_1 and D_2 and their associated bit lines B and $\bar{B}$.

Assuming that transistor T_1 is turned on and since the memory cell is a bistable element, it follows that transistor T_2 is off so that the drain of T_1 is held at approximately zero volts while the drain of T_2 is at the supply potential V_D. The cell is then storing a 0. For

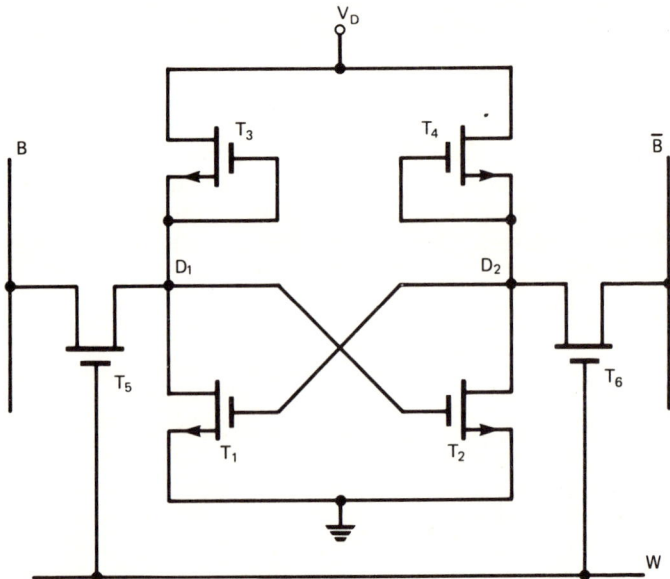

Figure 3.8 A static MOS RWM memory cell

reading, T_5 and T_6 are turned on by raising the word line potential; thus B is directly connected to approximately zero potential and $\bar{B}$ is held at the supply potential. The resulting voltage difference between B and $\bar{B}$ is then detected by a column sensing circuit.

For writing a data bit into the cell its true and complementary forms are transferred to the complementary bit lines B and $\bar{B}$. If a 1 is to be written into the cell, B is held at a potential of V_D while $\bar{B}$ is effectively grounded. The potential of the word line is now raised, turning on transistors T_5 and T_6 and thus connecting V_D to D_1 and zero volts to D_2. Hence the drain of T_1 is high and that of T_2 is low. This state is defined as a stored 1, which is maintained indefinitely until another write operation is initiated.

3.10 The dynamic RWM

The simplest type of memory cell consists of a transistor T and a capacitor C, as shown in Figure 3.9. The data is stored as charge on the capacitor C. When the capacitor is discharged to ground a 0 is stored in the cell by establishing the appropriate signal level on the bit line and then raising the potential of the word line so that the transistor conducts. When T conducts, the signal level on the bit line is transferred to the capacitor C.

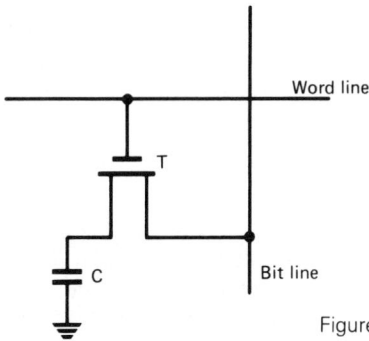

Figure 3.9 Single transistor dynamic RWM cell

The word and bit lines are initially charged to 0 and 12 volts respectively, and the read operation is performed as follows. To select a particular row, the potential of its word line is raised, thus turning on all the transistors in the row. When turned on, T behaves like a closed switch and the resulting circuit for one memory cell in the selected row is shown in Figure 3.10 where C_B represents the capacitance between bit line and ground.

If the voltage across C is initially zero, indicating the storage of a 0, there is a redistribution of charge between the two capacitors which results in the potential of the bit line being lowered. The fall in voltage is detected by a sensing circuit and corresponds to a stored 0.

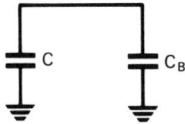

Figure 3.10 Connection of the data storage capacitance C to the bit line capacitance C_B when the transistor T conducts

If a 1 had been stored in the cell, capacitor C would have been charged to 12 volts, but due to leakage effects this potential will be less than 12 volts when transistor T is turned on. Again, there is a redistribution of charge when the two capacitors are connected through T which results in a lowering of the bit line potential but the magnitude of the fall is much less than in the case of a stored 1.

A major problem associated with this type of cell is that it is difficult to get a sufficiently large voltage swing to read the stored data due to the large bit line capacitance. Consequently the sense amplifiers are specially designed to ensure that they are adequately sensitive to detect the voltage changes. An additional problem is the loss of charge by capacitor C due to leakage effects. This will in time cause the voltage across C to fall below the threshold value for a stored 1 and it is necessary to refresh the cell at periodic intervals in order to restore the lost charge. This introduces an additional requirement for refresh circuitry when using this type of dynamic cell.

3.11 The bipolar RWM

The circuit configuration of the memory cell on a bipolar RWM chip shown in Figure 3.11 closely resembles that of the MOS RWM cell previously described. It consists of a bistable element in which the active devices are two multi-emitter transistors. To read out the information stored, the potential of the word line is raised and an indication of the true and complementary values of the stored data is transferred to the complementary bit lines. For writing, the potential of the word line is also raised, while the potential of one bit line is lowered, thus allowing the transistor associated with that bit line to conduct and, at the same time, turning the other transistor off.

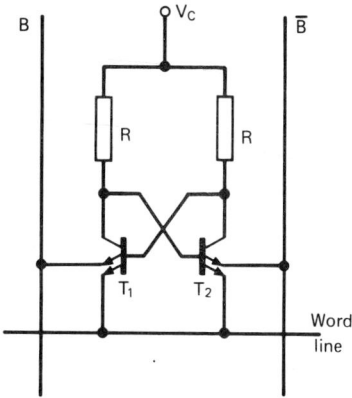

Figure 3.11 The bipolar static RWM cell

3.12 Memory expansion

It would be ideal if all the memory capacity required was available on one memory chip. In practice this is not always possible; for example, the required ROM capacity may be 64 8-bit words, i.e. 64 bytes, and there may only be chips with a capacity of 32 bytes available. To double the memory capacity two 32 byte memories can be connected in parallel, as illustrated in Figure 3.12.

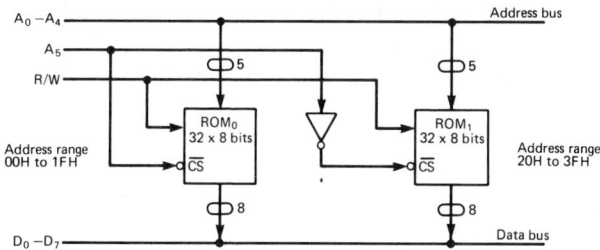

Figure 3.12 Expansion of memory capacity

The scheme shown requires six address lines. The least significant five address bits, A_0–A_4, are used for addressing the ROMs while the sixth address bit, A_5, generates the two chip select signals. For ROM_0 A_5 is taken directly to the chip select pin and for ROM_1 A_5 is inverted before being taken to the chip select pin. When $A_5 = 0$ ROM_0 is selected, and when $A_5 = 1$ ROM_1 is selected.

Each ROM stores 32 bytes and the address range for ROM_0, expressed in hexadecimal notation, is 00H to 1FH, while the address range for ROM_1, is 20H to 3FH. This gives an overall address range for the two ROMs of 00H to 3FH.

Besides the chip select signals each ROM is supplied with a read/write control signal R/W. Since both chips are ROMs only a read can take place from an addressed memory location. The read operation occurs when the chip select signal is low and R/W = 1. The data then appears at the eight output pins and is transferred to the common data bus.

If the required memory capacity had been 128 8-bit words then a 2-to-4 line decoder would be needed to generate the four chip select signals. The scheme is illustrated in Figure 3.13. The two address bits A_6 and A_5 supply the input signals to the decoder and the four output signals are used as chip select signals for each of the four ROMs. Each ROM is storing 32 8-bit words and the overall address range is 00H to 7FH.

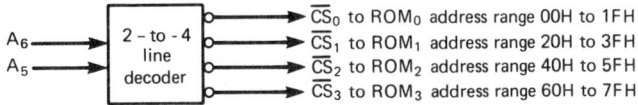

Figure 3.13 Chip select generation arrangements for memory of capacity 128 bytes consisting of four 32-byte ROMs

In some circumstances memory chips with a word length of eight bits may not be available. For example, the chips available may only have a storage capacity of 128 4-bit words while the microprocessor being used has a word length of eight bits. If two memory chips are connected, as illustrated in Figure 3.14, the available word length is extended from four bits to eight bits.

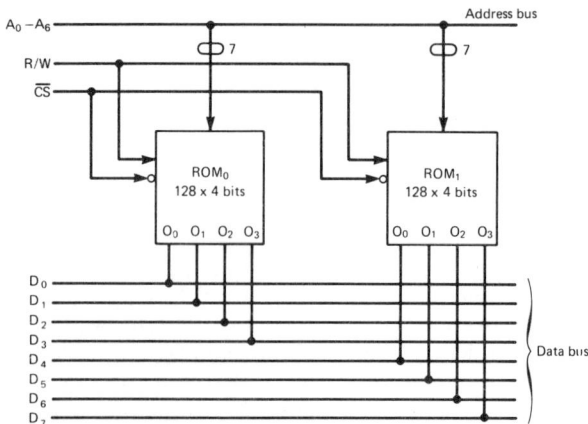

Figure 3.14 Word length expansion

The scheme shown requires two ROM chips ROM_0 and ROM_1, each of them addressed by address lines A_0–A_6 inclusive. The least significant four bits of any 8-bit word are stored in ROM_0 while the four most significant bits are stored in ROM_1. The required word is then accessed by applying the appropriate address code to the address lines and it appears at the outputs of the two ROMs for transfer to the data bus when the signals $\overline{CS} = 0$ and $R/W = 1$ are simultaneously applied to them. The address range of the stored data expressed in hexadecimal notation is 00H to 7FH.

3.13 Memory address decoding

An 8-bit microprocessor will normally have 16 address lines, so the maximum possible number of address locations is $2^{16} = 65\,536$. Frequently this amount of memory is not required in a particular application. Furthermore, the memory that is required normally consists of a number of chips which have to be selected via their chip select pins, and it is common practice to use some of the address lines for generating the chip select signals. If the memory required in a particular microprocessor application is 1 K byte of ROM and 1 K byte of RWM then a possible addressing scheme could be set up as shown in Figure 3.15.

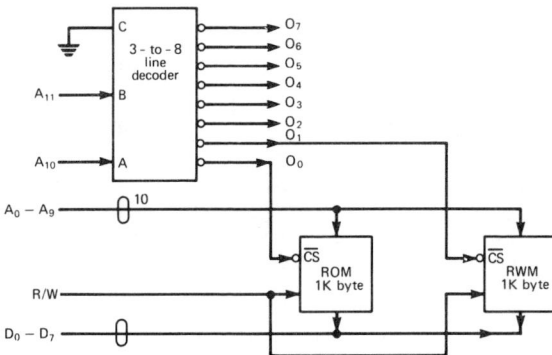

Figure 3.15 Non-absolute memory address decoding

The least significant ten address bits A_0–A_9 are used for addressing the 1024 memory locations in either ROM or RWM, depending upon which chip select signal is active low. The four most significant address bits A_{12}–A_{15} are left unconnected, while the two remaining address bits are used for generating the chip select signals. When $A_{10} = A_{11} = 0$, the output O_0 of the 3-to-8 line decoder is active low and provides the chip select signal for the ROM, and when $A_{11} = 0$ and $A_{10} = 1$ the output O_1 from the decoder is active low and is used as the chip select signal for the RWM. Six output lines of the decoder are not used but could, with some reorganization of the input signals, provide chip select signals for six other memory chips.

The address range for the ROM is 0000H to 03FFH, while the address range for the RWM is 0400H to 07FFH. If an address within the range 0000H to 07FFH is supplied by the microprocessor and the R/W control signal is high, the word stored at the addressed location will be transferred to the data bus.

The four most significant address lines can take up any one of 16 possible combinations of 0s and 1s and this implies that a specified memory location in either ROM or RWM can be accessed by 16 different addresses. In effect there is not a single absolute memory address for any specified memory location in either the ROM or RWM. This kind of addressing scheme is called non-absolute addressing.

The various combinations of the four most significant address bits are tabulated in Figure 3.16 and the corresponding address ranges for ROM and RWM are shown in the same table for each of these combinations. An address in any one of these ranges will access a memory location in one of the two memory chips.

Address bits				Address range ROM	Address range RWM	Address bits				Address range ROM	Address range RWM
A_{15}	A_{14}	A_{13}	A_{12}			A_{15}	A_{14}	A_{13}	A_{12}		
0	0	0	0	0000—03FF	0400—07FF	1	0	0	0	8000—83FF	8400—87FF
0	0	0	1	1000—13FF	1400—17FF	1	0	0	1	9000—93FF	9400—97FF
0	0	1	0	2000—23FF	2400—27FF	1	0	1	0	A000—A3FF	A400—A7FF
0	0	1	1	3000—33FF	3400—37FF	1	0	1	1	B000—B3FF	B400—B7FF
0	1	0	0	4000—43FF	4400—47FF	1	1	0	0	C000—C3FF	C400—C7FF
0	1	0	1	5000—53FF	5400—57FF	1	1	0	1	D000—D3FF	D400—D7FF
0	1	1	0	6000—63FF	6400—67FF	1	1	1	0	E000—E3FF	E400—E7FF
0	1	1	1	7000—73FF	7400—77FF	1	1	1	1	F000—F3FF	F400—F7FF

Figure 3.16 The 16 address ranges that will address the 2 K memory locations in ROM and RWM in the addressing scheme of Figure 3.15

3.14 Absolute addressing

Conversion of the above addressing scheme into an absolute addressing scheme, where a specified memory location in either ROM or RWM can only be accessed by one memory address, is shown in Figure 3.17. In this diagram the four most significant

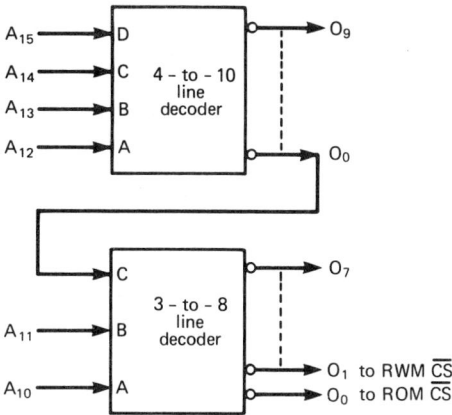

Figure 3.17 Absolute addressing using decoders

address bits are connected to the four inputs of the 4-to-10 line decoder. The only output from this decoder to be used is O_0 which is active low when $A = B = C = D = 0$. O_0 is then connected to the C input of the 3-to-8 line decoder. The outputs from this decoder which are used are O_0 and O_1 and the active low signals when they appear on either of these two lines are used to provide the chip select signals for the ROM and RWM. In effect the 4-to-10 line decoder is acting as an enabling chip for the 3-to-8 line decoder and the two memory chips can now only be addressed when $A_{15} = A_{14} = A_{13} = A_{12} = 0$.

Alternatively, a 4-bit comparator can be used as the enabling chip for a 3-to-8 line decoder which generates the chip select signals for a maximum of eight memory chips. A logic diagram illustrating this technique is shown in Figure 3.18.

All the Y inputs of the comparator are connected to ground while each of the four most significant address lines are connected to one of the X inputs of the comparator. When $A_{15} = A_{14} = A_{13} = A_{12} = 0$ the two input words X and Y are equal so that $X = Y$ and the comparator output is high. The output of the comparator is inverted and provides active low enable signals E_1 and E_2 for a 3-to-8 line decoder which has address lines A_9, A_{10} and A_{11} connected to its inputs. The decoder can provide eight active low outputs, each of which can be used as a chip select signal for a memory chip. Each of these memory chips can be addressed by the least significant nine address lines A_0–A_9 and hence they can have a maximum storage capacity of $1\,K$ byte each and an overall memory capacity of $8\,K$ bytes.

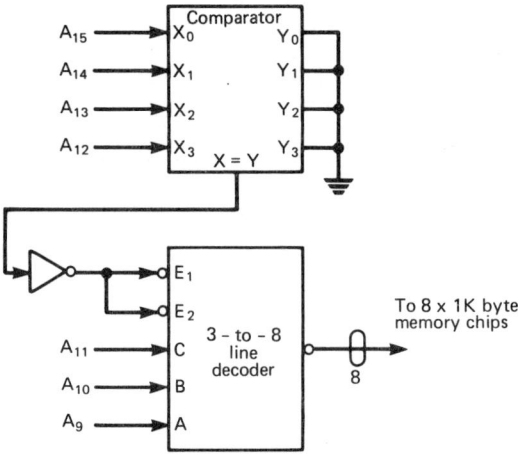

Figure 3.18 Absolute addressing using a comparator

Another method of absolute addressing involves a gate circuit for generating the decoder enabling signal. The method is illustrated in Figure 3.19. Providing $A_{15} = A_{14} = A_{13} = A_{12} = 0$ the enabling signals E_1 and E_2 are both active low, thuse enabling the 3-to-8 line decoder. As in the previous example, the decoder can select one of eight memory chips, each of maximum capacity 1 K byte, the selection being governed by the binary signals on the three address lines A_9, A_{10} and A_{11}.

Address decoding can also be achieved by using a PROM as a decoder. In the examples dealt with previously, decoders have

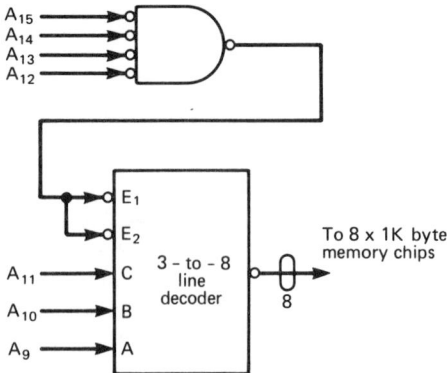

Figure 3.19 Absolute addressing using a logic gate

been used for address decoding and it will have been observed that each decoder output selects blocks of memory of the same size, but when a PROM is used as an address decoder the flexibility of selecting memory blocks of different sizes is available to the system designer. For example, consider the case of a micro-processor system that has the following memory requirements: 2 K bytes of RWM in the address range 0000h–07FFH, 4 K bytes of ROM in the address range 2000H–2FFFH, 4 K bytes of ROM in the address range 5000H–5FFFH, 2 K bytes of ROM in the address range 7000H–77FFH, and 16 K bytes of RAM in the address range C000H–FFFFH.

The memory requirements specified above are displayed on the memory map shown in Figure 3.20. This map covers the whole of the available addressable memory range from 0000H to FFFFH; that is, the 64 K addressable locations. The memory components specified for the microprocessor system in this example have less than 64K addressable locations, and furthermore the individual chips have dissimilar capacities, namely 2 K, 4 K and 16 K. The addressable locations on these chips occupy different regions of the memory map and, for the example quoted, the occupied regions are shown shaded. Vacant address space is represented by

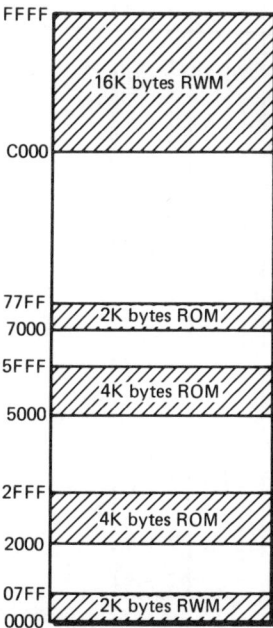

Figure 3.20 Memory map for the PROM decoder problem

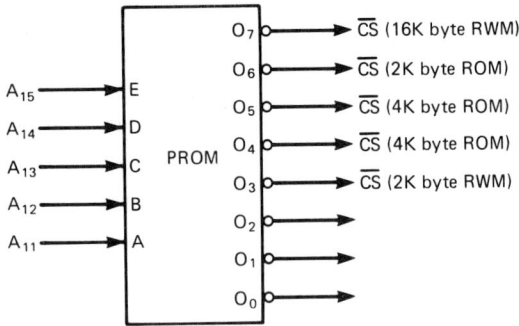

Figure 3.21 Connections for address decoding PROM

Address range	PROM address inputs					PROM outputs							
	E	D	C	B	A	O_7	O_6	O_5	O_4	O_3	O_2	O_1	O_0
0000—07FF	0	0	0	0	0	1	1	1	1	0	1	1	1
0800—0FFF	0	0	0	0	1	1	1	1	1	1	1	1	1
1000—17FF	0	0	0	1	0	1	1	1	1	1	1	1	1
1800—2FFF	0	0	0	1	1	1	1	1	1	1	1	1	1
2000—27FF	0	0	1	0	0	1	1	1	0	1	1	1	1
2800—2FFF	0	0	1	0	1	1	1	1	0	1	1	1	1
3000—37FF	0	0	1	1	0	1	1	1	1	1	1	1	1
3800—3FFF	0	0	1	1	1	1	1	1	1	1	1	1	1
4000—47FF	0	1	0	0	0	1	1	1	1	1	1	1	1
4800—4FFF	0	1	0	0	1	1	1	1	1	1	1	1	1
5000—57FF	0	1	0	1	0	1	1	0	1	1	1	1	1
5800—5FFF	0	1	0	1	1	1	1	0	1	1	1	1	1
6000—67FF	0	1	1	0	0	1	1	1	1	1	1	1	1
6800—6FFF	0	1	1	0	1	1	1	1	1	1	1	1	1
7000—77FF	0	1	1	1	0	1	0	1	1	1	1	1	1
7800—7FFF	0	1	1	1	1	1	1	1	1	1	1	1	1
8000—87FF	1	0	0	0	0	1	1	1	1	1	1	1	1
8800—8FFF	1	0	0	0	1	1	1	1	1	1	1	1	1
9000—97FF	1	0	0	1	0	1	1	1	1	1	1	1	1
9800—9FFF	1	0	0	1	1	1	1	1	1	1	1	1	1
A000—A7FF	1	0	1	0	0	1	1	1	1	1	1	1	1
A800—AFFF	1	0	1	0	1	1	1	1	1	1	1	1	1
B000—B7FF	1	0	1	1	0	1	1	1	1	1	1	1	1
B800—BFFF	1	0	1	1	1	1	1	1	1	1	1	1	1
C000—C7FF	1	1	0	0	0	0	1	1	1	1	1	1	1
C800—CFFF	1	1	0	0	1	0	1	1	1	1	1	1	1
D000—D7FF	1	1	0	1	0	0	1	1	1	1	1	1	1
D800—DFFF	1	1	0	1	1	0	1	1	1	1	1	1	1
E000—E7FF	1	1	1	0	0	0	1	1	1	1	1	1	1
E800—EFFF	1	1	1	0	1	0	1	1	1	1	1	1	1
F000—F7FF	1	1	1	1	0	0	1	1	1	1	1	1	1
F800—FFFF	1	1	1	1	1	0	1	1	1	1	1	1	1

Figure 3.22 Truth table for address decoding PROM

the unshaded regions and they could be used for other memory devices, or input/output devices if memory mapped I/O is being employed.

In this example the smallest size memory element is the 2 K byte ROM in the address range 7000H–77FFH and 11 address lines A_0–A_{10} are required for addressing memory locations on that chip. The remaining five address lines A_{11}–A_{15} can be used to address the decoding PROM, as illustrated in Figure 3.21.

Since the decoding PROM has five address inputs, a total of 32 8-bit words can be accessed by the five address lines, and each of these 32 locations can be identified with a 2 K block of address space, as shown at the left-hand side of the truth table in Figure 3.22. As there are only five memory devices to be selected just five of the PROM output lines are needed to provide the chip select signals for these devices.

3.15 Memory pages

In some microcomputing systems the 65 536 available addresses are divided into 256 blocks each containing 256 addresses, as illustrated in Figure 3.23. Each block is called a *page*. The page number is identified by the two most significant hexadecimal digits in the overall address, while the two least significant hexadecimal

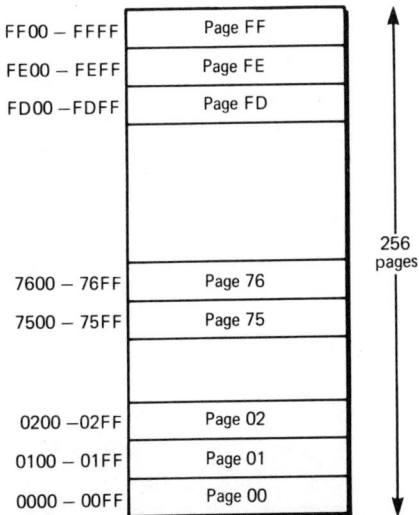

FF00 – FFFF	Page FF
FE00 – FEFF	Page FE
FD00 –FDFF	Page FD
7600 – 76FF	Page 76
7500 – 75FF	Page 75
0200 –02FF	Page 02
0100 – 01FF	Page 01
0000 – 00FF	Page 00

256 pages

Figure 3.23 The division of the total address space into memory pages

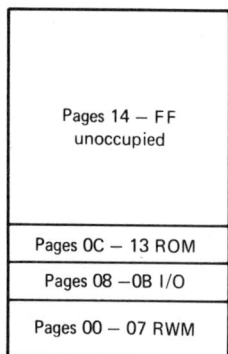

Figure 3.24 A page-organized addressing scheme

digits will identify an address on the page. For example, if the address is 1E66 then the page is identified by the hexadecimal digits 1E and the address on the page is identified by the digits 66.

The use of paging leads to page-organized addressing schemes such as the one shown in Figure 3.24. In this scheme the first eight pages are allocated to RWM, the next four pages are reserved for I/O operations, and the next eight pages are allocated to ROM, leaving the last 236 pages vacant.

Problems

3.1 The memory capacity of a given ROM is 2 K bytes. How many data input lines and data output lines are associated with this chip? Draw a block diagram of the chip assuming (a) one-dimensional addressing and (b) two-dimensional addressing, and compare the two methods of addressing. What is the hexadecimal equivalent of the address $(779)_{10}$?

3.2 Determine the size of a ROM which would be required:

(a) to convert a two-digit BCD number into its binary equivalent,
(b) to compare two 4-bit words,
(c) to implement a binary adder/subtractor having a mode control m to distinguish between addition and subtraction and capable of handling 4-bit numbers.

3.3 How many 256 × 8 ROM chips are required to provide a memory capacity of 2 K bytes? Draw a block diagram showing the addressing arrangements assuming that a decoder is available for providing the chip select signals. Draw a memory map for the system and determine how much unused memory capacity is available for RWM, assuming a total of 16 address lines in the system.

3.4   A microcomputer memory system requires 1 K byte of RWM and 1 K byte of ROM. Available are 256 × 8 RWM chips having two chip select pins, one active low and the other active high, and 512 × 8 ROM chips having one chip select pin that requires an active low signal. Additionally, a 3-to-8 line decoder requiring two active low and one active high enable signal is available. Draw a block diagram for the system, assuming that there are 16 address lines. List the memory map and give the address ranges of the system in hexadecimal.

Extend the system you have devised to include an additional 1 K byte of RWM and 2 K bytes of ROM. Draw the complete memory map and indicate the size of decoders required.

3.5   Explain the difference between absolute and non-absolute address decoding. Give examples of absolute address decoding using (a) decoders (b) comparators and (c) logic gates.

3.6   A microcomputer system having 16 address lines requires 2 K bytes of ROM and 2 K bytes of RWM. 1 K byte ROM and RWM chips are available for implementing the memory requirements. A 3-to-8 line decoder provides non-absolute memory decoding. Determine the address ranges that will access the 1024 address locations in each memory chip.

3.7   A multiplier for multiplying two 8-bit numbers is to be used in conjunction with an 8-bit microprocessor. Design a suitable multiplier system using 256 × 8 bit ROMs and 4-bit adders.

4 Small computer architecture

4.1 Introduction

The microprocessor is the most important component in a microcomputing system. It is the central processing unit (CPU) of the microcomputer and it is in this unit that all arithmetic and logic operations are performed on data which has been brought there either from memory, or from the outside world via an input port. Additionally, when data has to be transferred to the outside world it is normally despatched there via the microprocessor and an output port. The microprocessor receives its instructions for performing these various operations from the program stored in ROM and it then operates on data in accordance with the received instructions.

In the early stages of this chapter the component parts of a microcomputer, the method of their interconnection and their functions will be described. This will be followed by a discussion of the internal architecture of a microprocessor, outlining the purpose and operation of the component parts of the device. In particular, the internal architecture and operation of an 8-bit microprocessor, the Intel 8085A, will be examined in some detail.

4.2 The block diagram of a microcomputer system

A generalized block diagram of a basic microcomputer system is shown in Figure 4.1. This diagram identifies four main blocks.

(1) The CPU or microprocessor, which controls the addressing of memory and input/output, receives data from RWM or the outside world, processes it, and then returns it to a location in memory or to the outside world. Additionally, it receives instructions from ROM which it first decodes and then implements. Finally, the microprocessor is the source of the system control which allows interaction between the various

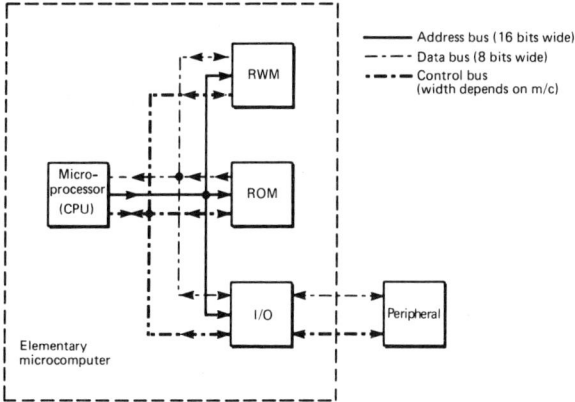

Figure 4.1 Block diagram of a basic microcomputing system

parts of the computer system and also the system timing which is provided by the clock.

(2) ROM is used to store the program and, as the name implies, information can only be read from this type of memory element. The information in this case is the program, which consists of a sequence of instructions representing the algorithms required to match the microcomputer to its application. The digital engineer designs the algorithms once and then stores them in ROM, thus dedicating the micro-computer to a particular application.

(3) RWM is generally used for storing data which may be obtained from the outside world via the input ports and transferred there from the microprocessor after an arithmetic or a logical operation. Alternatively, instructions could be stored in RWM during, for example, the development of programs. In comparison to ROM, RWM is volatile and power failure will result in the disappearance of its contents. It is for this reason that programs are preferably stored in non-volatile ROM.

(4) The I/O block provides the main path for the transfer of data to and from the outside world. It may consist of a simple I/O port or, alternatively, a more complex programmable input/out chip (PIO). The function of the I/O block is twofold:
 (a) to maintain input data sufficiently long in order that it can be read by the microprocessor, or
 (b) to maintain output data long enough to enable it to be read by a peripheral.

The electronic device within the I/O block of an 8-bit machine for maintaining the data in this way is an 8-bit latch.

The peripheral shown in Figure 4.1 is the device by means of which the user or application communicates with the basic microcomputer. Typical peripherals in use with a small computer of this type are:

(a) Keyboard
(b) Teletype
(c) Visual display unit (VDU)
(d) A/D and D/A converters
(e) Universal synchronous/asynchronous receiver/transmitter (USART)
(f) Floppy disk.

The interconnection of the peripheral and the computer is achieved via an interface whose function is to receive status signals from both the computer and the peripheral and then generate the appropriate command signals for the two devices in their correct sequence.

4.3 The bus system

The various elements of the basic microcomputer, illustrated in Figure 4.1, are interconnected via the common bus system which is characterized by three distinct component parts:

(1) The unidirectional address bus, which originates from the microprocessor. Commonly, the address bus is 16 bits wide, and this allows access to a maximum of $2^{16} = 65\,536$ memory locations. In practice, the address bus serves two purposes. First, it is used to select a particular device such as a ROM or RWM chip, and secondly, it is used for addressing a particular memory location within the selected chip.
(2) The bidirectional data bus, which in the 8-bit machine is eight bits wide. It is used for carrying data in both directions between memory and the microprocessor and also between the outside world and the microprocessor. Additionally, instructions are transferred from ROM to the microprocessor (μP) on the data bus.
(3) The control bus carries the signals which are used to synchronize the activities of the various elements in the computer system. The bus is bidirectional since synchronization signals can be carried to and from the μP and its width

varies from machine to machine. There are some control signals such as READ, WRITE and INTERRUPT which are common to all machines, but there are others which are peculiar to a particular make of machine.

In some machines the data bus is time-multiplexed with eight lines of the address bus. This implies that these eight lines can carry both data and address information during separate time intervals which are carefully defined by the system clock. This particular technique is employed in the Intel 8085A and it has the obvious advantage of releasing pins on the microprocessor chip and thus making them available for other purposes.

The common bidirectional data bus may be regarded as a highway for data flow. Apart from a few exceptions, each device in the system has to be connected to the data highway, irrespective of whether it is internal to the block diagram shown in Figure 4.1, for example RWM or a peripheral such as an A/D converter. It is also clear that at any given instant of time it is only possible for a single device to be using the data highway. As a consequence, it is normal procedure to isolate devices from the data bus by tri-state buffers, where a single tri-state buffer is connected to each data line. When disabled, the tri-state buffer presents a high impedance to the data bus, thus isolating the data line from the common data highway, and at the same time preventing the transfer of data to the highway.

Tri-state buffers also serve another function. The input of a peripheral or a memory chip represents a load on the output driving it. Typically, a MOS microprocessor lacks the output drive needed and, because of this buffers or line drivers are used to boost the drive capability of the bus system. Generally speaking a buffer will provide 30 to 40 mA of drive current, and hence is capable of driving 18 to 24 standard TTL loads.

4.4 Microprocessor architecture

The microcomputer system shown in Figure 4.1 may be regarded as a collection of addressable registers, and data is moved between registers along the data bus. The collection of registers constituting the system and the possible transfers between them comprises the system architecture.

Although the microprocessor is an element in the overall system, it possesses its own unique architecture. Internally, it consists of a selection of registers and associated logic elements on the microprocessor chip, and data is transferred between these registers via the internal data bus. A diagram illustrating the typical architecture for a microprocessor is shown in Figure 4.2.

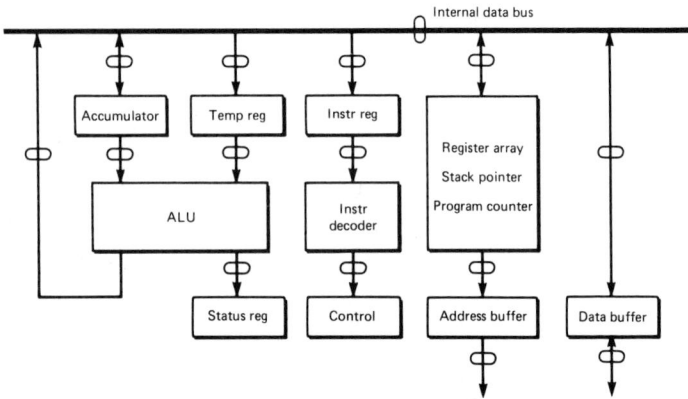

Figure 4.2 Basic microprocessor architecture

The microprocessor receives instructions via the data bus from memory. Additionally, it receives data either from memory or from the outside world. Information is received in an ordered sequence and this is achieved by storing the program in consecutive locations in memory. Operations resulting in the transformation or transfer of received data are implemented by executing the instructions received from memory.

4.5 Internal registers

The *program counter* (PC) is a 16-bit register which is connected to the 16-bit address bus via an address buffer and counts when it is clocked. The presence of the program counter is fundamental to the execution of a program in the sense that it accesses the memory where the program is stored in successive locations. As an example of the operation, the contents of the program counter in Figure 4.3, 087AH, are output onto the address bus via the address buffer. The contents of the addressed location 32 H are placed on the system data bus a short time after a read control signal is received by the program memory from the microprocessor. This time delay is the memory access time.

The contents of 087AH may be either an instruction opcode or data; in either case the information on the data bus is transferred

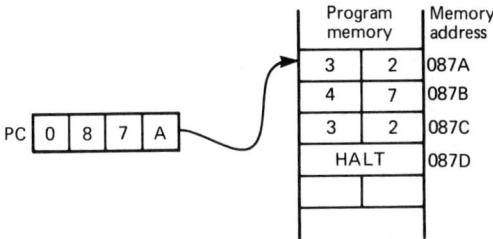

Figure 4.3 Addressing memory with the program counter

from memory to the appropriate register in the microprocessor via the data buffer. After the transfer has taken place PC is incremented by a signal from the machine control unit so that it is now addressing memory location 087BH. Sequential operation of this kind continues until such time as either a halt instruction, HLT, or a jump instruction, JMP, appears in the main program sequence. In the case of the halt instruction PC stops counting, while with the jump instruction the contents of PC are replaced by a non-consecutive address.

A second internal register to be found in all microprocessors is the *instruction register* (IR) which for an 8-bit machine is eight bits wide. An instruction stored in program memory may consist of one, two or three bytes and the first byte of all instructions is the instruction opcode, which defines the operation to be performed by the machine. Typical instruction opcodes define operations such as Load, Store, and Add, etc. The output of IR is taken to the instruction decoder, as illustrated in Figure 4.4, where the instruction opcode, in this case 32H, is interpreted by the decoder logic circuit before instruction execution begins.

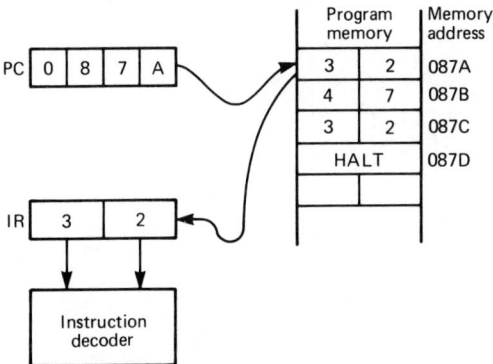

Figure 4.4 Transfer of an instruction opcode to the instruction decoder via the instruction register

A third important internal register in an 8-bit machine is the 8-bit wide *accumulator*, which alway contains one of the operands to be transferred to the arithmetic/logic unit when the machine is instructed to perform either an arithmetic or logic operation. The second operand for the operation is brought to a temporary register. Both the accumulator and the temporary register have direct access to the ALU input and, after an arithmetic or logic operation has been performed in it, the result is returned to the accumulator via the internal data bus of the microprocessor. Since one of the operands for a given operation is always held in the accumulator, the microprocessor is called a single address machine, the address for the other operand always being specified. Additionally, for all normal I/O operations input and output data is routed through the accumulator, hence it is clear that it is one of the key registers in the machine.

It is common practice to reserve a section of the RWM for *stack* operations. The function of the stack is to store the contents of the microprocessor registers during a *subroutine* or an *interrupt* on a last-in, first-out (LIFO) basis. This means that the stack is organized in such a way that information stored on it is read out in reverse order to the order in which it was written into it.

The contents of the stack are accessed by a 16-bit register called the *stack pointer* which has to be initialized at the beginning of the main program sequence so that it is pointing to the memory location adjacent and prior to the first stack location, or alternatively, the first stack location to be written into, depending upon the convention employed by the microprocessor being used. This point is illustrated in Figure 4.5.

Data is moved onto and off the stack with the aid of two instructions. The instruction PUSH enters information onto the

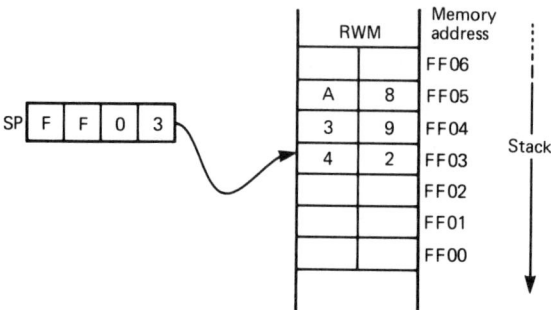

Figure 4.5 Stack pointer pointing at last location written into

stack, while the instruction POP releases data in the reverse order from the top of the stack. The use of these instructions for controlling stack operations will be dealt with more fully in a later chapter.

Some microprocessors include an *index register*, which is used when the machine is operated in the indexed addressing mode. When this mode of addressing is employed, the effective operand address is obtained by adding the contents of the index register to the base address which is provided by the address portion of the current instruction. This kind of addressing technique is very useful when handling data tables stored in memory.

Besides the registers that perform specific functions there are frequently temporary registers and general purpose storage registers available within the microprocessor. For example, in the Intel 8085A there are two temporary registers available, identified by the letters W and Z, which are not accessible to the programmer.

Some three-byte instructions contain a two-byte address in the last two bytes of the instruction. When these two bytes are fetched from program memory they are provided with a temporary home in registers W and Z, and during the instruction execution the address stored in W and Z is transferred to the memory address buffer and thence to the address bus.

The general storage registers can be used for a variety of functions. They may be used as a temporary store for intermediate results, thus eliminating a transfer of this data to memory and returning it to the microprocessor when required, hence speeding up machine operation and program execution. Also they can be used to store data that is frequently used during program execution, or in the absence of an index register one of them can be utilized for indexed addressing. In some programs a loop has to be repeated a specified number of times. The loop count can be stored in a general purpose register that is decremented every time the loop is traversed; when the count is zero the machine can continue with the main program sequence.

Additionally, general storage registers can be used as a pointer to an address in memory to or from which data has to be transferred. The transfer can be from microprocessor to memory, or vice versa. A register pair can be loaded with the memory address and a subsequent instruction can be used to transfer data to or from that address. After the transfer has been effected the contents of the register pair can be either incremented or decremented, thus giving access to the next consecutive memory location.

4.6 The arithmetic/logic unit

It is in this unit that all the arithmetic and logical operations available to the microprocessor are implemented. The connections to the ALU are shown in Figure 4.6. One of the operands will always be found in the accumulator since it has been inserted there after the previous operation or it has been brought there from memory. The second operand is transferred to the temporary register via the internal data bus. Selection of the required operation is achieved via the function select inputs and after the operation has been performed the result is returned to the accumulator via the internal data bus.

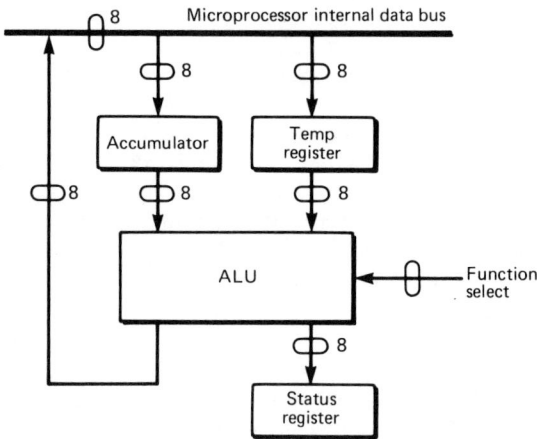

Figure 4.6 Schematic diagram of the ALU and its associated registers

Typical examples of the operations that can be performed by the ALU on data are:

(1) Add
(2) Subtract
(3) AND, OR and XOR
(4) Shift right or left
(5) Complement.

Figure 4.6 also reveals the presence of another internal register, called the *status register*, which is best dealt with in the context of ALU operations. After an arithmetic or a logical operation has been performed, a decision can be made by the microprocessor which depends upon the result of this operation. For example, if the result of such an operation is that the contents of the accumulator are zero then a jump may be required, based on this condition, to some new memory location which may be either forward or backward in the main program sequence. If the contents of the accumulator are not zero the jump will not take place and the next consecutive instruction will be implemented. This type of decision-making instruction is available in the machine instruction set; for example, 'jump on zero' (JZ) is one such conditional instruction. Hence the condition of the various bits stored in the status register may be examined after the implementation of an arithmetic or logic operation to decide the address of the next instruction and, consequently, the desired program sequence. The bits stored in the status register are normally referred to as flags, and typical examples of status flags are Carry, Zero, Parity, Sign and Overflow.

4.7 Basic microprocessor operation

The operation of any microprocessor can be divided into two parts, the *instruction fetch* and the *instruction execute*. During the instruction fetch, a series of data transfers from memory to microprocessor are performed until the whole of the instruction has been transferred. After the whole instruction has been received by the microprocessor, instruction execution can take place. These operations can be described by the state diagram shown in Figure 4.7, which consists of three states, FETCH, EXECUTE and HALT. Transition from FETCH to EXECUTE

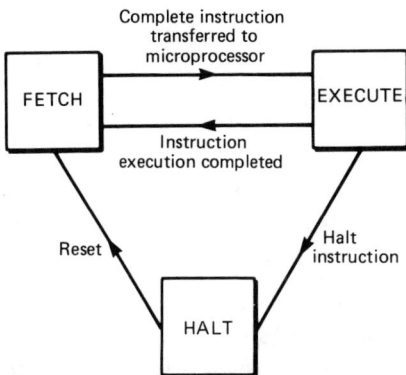

Figure 4.7 The basic operation of a microprocessor in terms of the FETCH, EXECUTE and HALT states

occurs when the whole of the instruction has been transferred to the microprocessor. Execution can now take place and a transition back to the FETCH state only occurs when the execution of the instruction has been completed. Hence, when implementing a program the operation consists of a sequence of fetch and execute cycles controlled by the microprocessor control unit.

If, after the execution of an instruction, the next instruction in the program sequence is HLT, a transition is made to the HALT state. The machine will remain in this state until a reset signal is received which returns the machine to the FETCH state.

There are certain basic control signals which have to be generated within the control unit of any microprocessor in order to initiate the operations which take place in the fetch and execute cycles. The two most important of these are the read ($\overline{\text{RD}}$) and write ($\overline{\text{WR}}$) signals which are used for reading from memory or an input port and for writing into memory or an output port. In some cases a more precise set of signals may be required which are designed to perform exactly the same function but distinguish between memory and I/O reads and also between memory and I/O writes. They are memory read ($\overline{\text{MEMR}}$), memory write ($\overline{\text{MEMW}}$), I/O read ($\overline{\text{I/OR}}$) and I/O write ($\overline{\text{I/OW}}$). This second set of signals can be generated from the $\overline{\text{RD}}$ and $\overline{\text{WR}}$ signals and a status signal IO/$\overline{\text{M}}$, which differentiates between I/O and memory operations. The signal IO/$\overline{\text{M}}$ will be logical 1 during an I/O operation and logical O during a transfer either to or from memory. A circuit for the generation of one of these signals is illustrated in Figure 4.8. It is left to the reader to develop the gate circuits required for the remaining three signals.

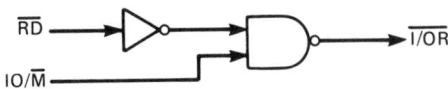

Figure 4.8 Generation of the $\overline{\text{I/OR}}$ signal

Besides the generation of these basic signals, the control section of a microprocessor receives control signals from peripherals and generates acknowledging control signals to be returned to the peripheral. Two typical examples of received control signals are HLD (Hold) and INT (Interrupt). The hold signal is used to initiate a *direct memory access* (DMA) transfer between a peripheral and memory, while the interrupt signal temporarily stops the main program sequence and the machine enters an *interrupt service routine* (ISR) at the initiation of the interrupting peripheral. In response to these received control signals the microprocessor generates acknowledging signals such as HLDA (Hold Acknowledge) and INTA (Interrupt Acknowledge).

4.8 The 8085A microprocessor

The Intel 8085A is a 5 V, 8-bit microprocessor manufactured on a single 40-pin chip. Various technologies have been used in the fabrication of microprocessors. First generation devices used PMOS technology, but the 8085A is a second generation microprocessor and NMOS technology has been used in its fabrication. The 8085A chip is shown in Figure 4.9 and the function of each pin is identified by its associated label.

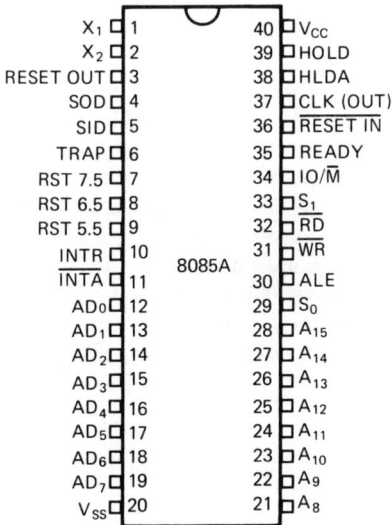

```
        X₁ ▯  1        40 ▯ V_CC
        X₂ ▯  2        39 ▯ HOLD
 RESET OUT ▯  3        38 ▯ HLDA
       SOD ▯  4        37 ▯ CLK (OUT)
       SID ▯  5        36 ▯ RESET IN
      TRAP ▯  6        35 ▯ READY
   RST 7.5 ▯  7        34 ▯ IO/M̄
   RST 6.5 ▯  8        33 ▯ S₁
   RST 5.5 ▯  9        32 ▯ R̄D̄
      INTR ▯ 10        31 ▯ W̄R̄
      ĪNTA ▯ 11  8085A 30 ▯ ALE
      AD₀ ▯ 12        29 ▯ S₀
      AD₁ ▯ 13        28 ▯ A₁₅
      AD₂ ▯ 14        27 ▯ A₁₄
      AD₃ ▯ 15        26 ▯ A₁₃
      AD₄ ▯ 16        25 ▯ A₁₂
      AD₅ ▯ 17        24 ▯ A₁₁
      AD₆ ▯ 18        23 ▯ A₁₀
      AD₇ ▯ 19        22 ▯ A₉
      V_SS ▯ 20        21 ▯ A₈
```

Figure 4.9 Pin connections for the 8085A

All instruction opcodes for operations such as load, add, etc. are restricted to eight bits, and consequently there can only be a maximum of 256 instructions. In practice there are only 74 distinct instructions, but some instructions such as ADD appear in the instruction set in a variety of different forms, and hence the majority of the 256 possible combinations of eight bits are utilized.

A microprocessor is a synchronous sequential machine. It performs its operations sequentially, and these operations are synchronized to the microprocessor clock. The frequency-determining elements for the clock are connected between the terminals marked X_1 and X_2 in Figure 4.9 and they can be either a crystal or an RC network connected as shown in Figures 4.10(a) and (b), or, alternatively, it is possible to connect an external clock across the two terminals. The internal clock frequency is half the

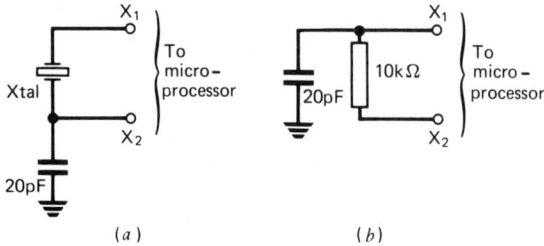

Figure 4.10 (a) Crystal controlled clock (b) RC controlled clock

frequency of oscillation. If, for example, the crystal frequency is 6 MHz then the internal clock frequency will be 3 MHz. A clock output pin is provided so that the microprocessor clock frequency can be distributed to all parts of the microcomputer system.

The microprocessor can be initialized by the $\overline{\text{RESET IN}}$ signal. When this signal is low the program counter contents are returned to all the zeros. When $\overline{\text{RESET IN}}$ becomes high again the contents of the program counter are transferred to the address bus. Program execution will then commence with the contents of memory location 0000H, which should contain the first instruction opcode. A manual reset circuit is shown in Figure 4.11.

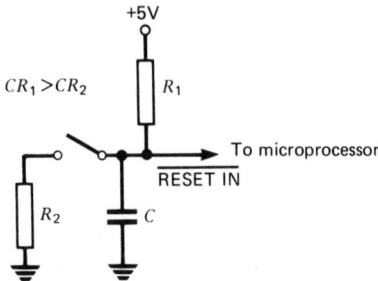

Figure 4.11 Combined manual reset and automatic power-on circuit

The microprocessor also provides a RESET OUT signal, which can be used to indicate that the machine is being reset and hence can perform a useful function as a system reset signal.

The 8085A has 16 address lines, and thus the machine is capable of accessing 65 536 memory locations. The eight most significant address lines A_8–A_{15} are uniquely used for addressing memory. However, the least significant eight lines AD_0–AD_7 serve a dual function in that they are used for both address and data. Time multiplexing for these eight lines is established by means of a control signal ALE (address latch enable) which is generated in the control section of the 8085A. This signal latches the eight address bits on lines AD_0–AD_7 into an on-chip latch or holding register on the memory chip. If the chip being addressed does not contain such a register an external one must be provided and the eight least significant address bits are latched into this register by the ALE signal. After latching, lines AD_0–AD_7 are free and can be used for the transmission of data.

Besides addressing memory, the least significant eight address lines can be used for addressing I/O ports. Since eight bits are available for I/O addressing, a maximum number of 256 input ports and 256 output ports can be addressed. The address of the port is contained in an input (IN) or output (OUT) instruction which also implements the transfer of data from the port to the accumulator, or vice versa. In practice, the port address also appears on the eight most significant lines of the address bus.

During a memory reference the microprocessor generates read ($\overline{RD}$) and write ($\overline{WR}$) control signals. In the earlier chapter on Memory it was shown that these two signals can be combined with a chip enable signal ($\overline{CE}$) so that the data can be read from or written into memory.

At the beginning of each instruction cycle the microprocessor generates a status signal, IO/$\overline{M}$ which distinguishes between a reference to memory (IO/$\overline{M}$ = 0) or a reference to an I/O port (IO/$\overline{M}$ = 1). If this signal is combined with $\overline{RD}$ or $\overline{WR}$, four separate control signals $\overline{MEMR}$, $\overline{MEMW}$, $\overline{I/OR}$ and $\overline{I/OW}$ can be derived, which will distinguish specifically between memory and I/O references.

The 8085A is also designed to receive a READY signal, which is used by a peripheral communicating with the microprocessor to indicate that data is available for transfer. If the READY signal is high during a read or write operation it indicates that a data transfer can take place. If, however, the READY signal is low then the microprocessor enters a WAIT state and remains there for an integral number of microprocessor clock cycles until READY goes high before completing the read or write operations.

The previous paragraphs have dealt with those microprocessor pins associated with switching on, clock generation and transfers from memory and I/O to the microprocessor, and vice versa. The function of the remaining pins on the microprocessor chip not covered in these paragraphs will be developed later in this chapter or in succeeding chapters as the need arises.

4.9 The architecture of the 8085A

The architecture of the 8085A is essentially the same as the basic microprocessor architecture shown in Figure 4.2. A more comprehensive diagram of the 8085A architecture provided by the manufacturer is given in Figure 4.12. This diagram provides the additional details listed below:

(1) A more comprehensive description of the general purpose register array, which identifies six 8-bit registers B, C, D, E, H and L which can be used singly or as a 16-bit register pair.
(2) Information regarding the input and output signals to and from the control unit. In particular, it identifies the incoming (HLD) and outgoing (HLDA) *direct memory access* (DMA) control signals and also the two status signals S_0 and S_1 which, when examined with IO/$\overline{M}$, defines the machine cycle presently being performed by the machine.
(3) A 5-bit register labelled flag flip-flops, which corresponds to the status register in Figure 4.2 and provides the decision-making information required by the machine.
(4) Serial I/O register and an interrupt control register, whose functions will be described in more detail later.

4.10 The operation of the 8085A

During the execution of a program by the 8085A a series of instruction fetches and instruction executes are performed sequentially. In the course of the instruction fetch the total instruction, which may consist of one, two or three bytes, is brought to the microprocessor. When the whole instruction has reached the microprocessor, instruction execution begins. The total time needed to fetch and execute the instruction is called the *instruction cycle time*.

Each reference the microprocessor makes to memory or an I/O port, irrespective of whether it is a read or write, is called a *machine cycle*, and clearly an instruction cycle will consist of a

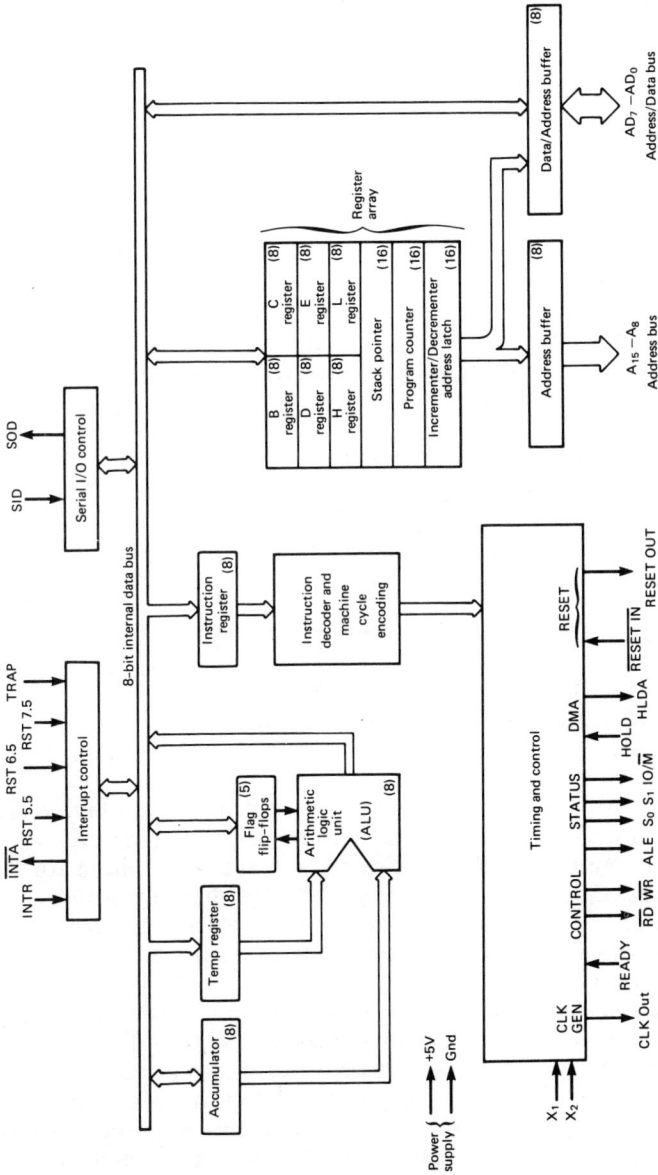

Figure 4.12 8085A CPU functional block diagram

series of machine cycles. In all, the 8085A has seven distinct machine cycles, and they are:

1. OPCODE FETCH   (OF)
2. MEMORY READ   (MR)
3. MEMORY WRITE   (MW)
4. I/O READ   (IOR)
5. I/O WRITE   (IOW)
6. INTERRUPT ACKNOWLEDGE   (INA)
7. BUS IDLE   (BI)

Each of these machine cycles is defined by a particular combination of the status signals $IO/\overline{M}$, S_0 and S_1, as illustrated in the tabulation given in Figure 4.13. The logical values of the control signals are also tabulated in Figure 4.13 and it should be observed that for any read operation $\overline{RD}$ is low, and for any write operation $\overline{WR}$ is low.

Machine cycle	Status signals			Control signals		
	$IO/\overline{M}$	S_1	S_0	$\overline{RD}$	$\overline{WR}$	$\overline{INTA}$
OF	0	1	1	0	1	1
MR	0	1	0	0	1	1
MW	0	0	1	1	0	1
IOR	1	1	0	0	1	1
IOW	1	0	1	1	0	1
INA	1	1	1	1	1	0
BI DAD	0	1	0	1	1	1
INA/RST/TRAP	1	1	1	1	1	1
HALT	T.S	0	0	T.S	T.S	1

Figure 4.13 The machine cycle chart for the 8085A. TS = High impedance

No instruction cycle consists of more than five machine cycles. Most instructions require one, two or three machine cycles, and only a limited number require four or five. Since an instruction consists of one, two or three bytes, the number of machine cycles required for an instruction fetch is between one and three. Any remaining machine cycles after the instruction fetch comprise the instruction execution.

Each machine cycle consists of a number of T-states, where the time duration of such a state is equal to the periodic time of the microprocessor clock. In practice, most machine cycles consist of three T-states, with the exception of the opcode fetch which requires between four and six T-states. If the microprocessor clock frequency and the number of T-states associated with an instruction are known, it is a simple matter to determine the time duration of the instruction cycle. For example, the instruction

LDA (Load Accumulator Direct) requires four machine cycles and 13 T-states. For a microprocessor clock frequency of 3 MHz, the time taken to implement the instruction is

$$t_I = 13 \times \frac{1}{3 \times 10^6} \text{ s}$$
$$= 4.33 \, \mu\text{s}$$

The timing of the LDA instruction is illustrated in Figure 4.14(*a*). LDA is the mnemonic representation of the instruction and may be interpreted as 'Load the accumulator with the contents of the memory location specified by bytes 2 and 3 of the instruction.' Thus the three bytes of the instruction are stored in the program memory as shown in Figure 4.14(*b*).

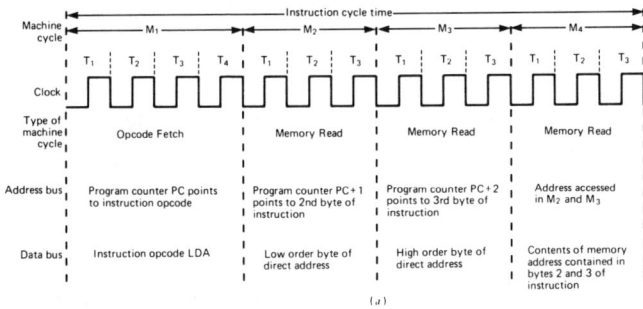

(*a*)

(*b*)

Figure 4.14 (*a*) Timing diagram for the LDA instruction (*b*) Storage of the LDA instruction in program memory

This instruction requires four machine cycles to complete, and these are detailed below:

Instruction byte	Machine cycle		States
LDA { Opcode	M₁ Opcode Fetch	Instruction Fetch	4
Addr. Low	M₂ Memory Read		3
Addr. High	M₃ Memory Read		3
	M₄ Memory Read	Instruction Execute	3

During the machine cycle M_1, the contents of the program counter are placed on the address bus and a memory read is performed by the microprocessor, the instruction opcode being transferred to the instruction register. The next two machine cycles M_2 and M_3 are both memory reads and in each of these cycles one byte of the address of the data to be loaded into the accumulator is transferred to the temporary registers W and Z in the microprocessor. The instruction fetch is complete, and the instruction execution can now take place. The contents of registers W and Z are placed on the address bus and the contents of the addressed location are transferred from memory to accumulator.

4.11 State and timing diagrams for the 8085A

Since the 8085A is a synchronous sequential state machine its behaviour can be represented by a state diagram such as the one shown in Figure 4.15. Machine states are represented by squares, the lines between the states represent transitions, and the arrowheads give the direction of the transition. Signals initiating transitions are placed by the arrowheads.

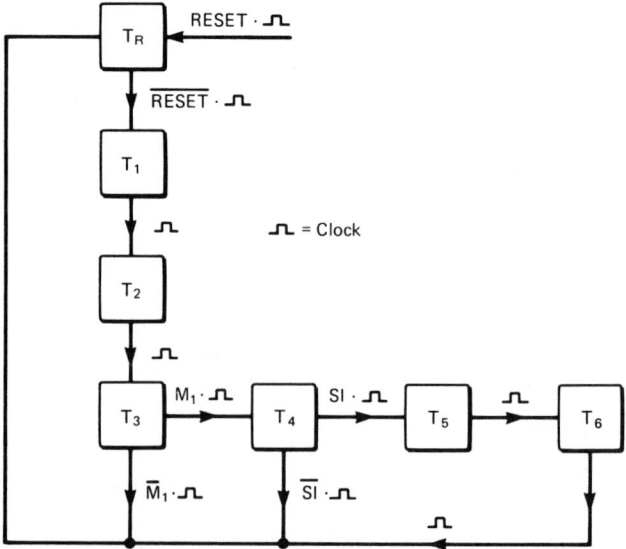

Figure 4.15 Basic state diagram for the 8085A

Certain events are associated with each of the machine states. They are:

T_1 (a) Memory or I/O device address placed on the address bus.

       (b) ALE signal generated.

       (c) Status information defining type of machine cycle appears on S_0, S_1 and IO/$\overline{M}$.

T_2 (a) READY signal sampled.

       (b) PC is incremented if machine cycle is part of instruction fetch.

T_3 Opcode or data byte transferred to/from microprocessor.

T_4 Contents of instruction register decoded.

T_5 and T_6 Used to complete execution of some instructions.

There are some instructions such as DCX rp (decrement register pair) which are executed in the opcode fetch machine cycle, and for this reason two extra states T_5 and T_6 are provided. This has been indicated on the state diagram in Figure 4.15 by using the combination of the transition signal SI and the clock, SI.⎍ to take the machine from T_5 to T_6, where SI indicates the presence of an instruction requiring a six-state opcode.

To enter the four- or six-state opcode fetch cycle, a transition signal M_1 has to be generated when the machine is in state T_3 and the combination of this signal with the clock M_1.⎍ takes the machine from T_3 to T_4. If, when in state T_4, the signal SI has not been generated, the machine returns directly to T_1 and has performed the normal four-state opcode fetch. For the second machine cycle of an instruction, the signal M_1 is not generated in T_3 and the machine returns directly to T_1 from T_3. This would represent the typical three-state cycle the machine would traverse when performing a memory read or memory write.

The timing diagram for the DCX rp instruction is shown in Figure 4.16. Status signals $S_0 = S_1 = 1$ and IO/$\overline{M} = 0$ generated in T_1 define the type of machine cycle, in this case an opcode fetch. Address information placed on address lines A_8–A_{15} in state T_1 remains there until at least state T_4, while the address information placed on lines AD_0–AD_7 remains there for one clock cycle only and is latched by the ALE signal generated in T_1. The $\overline{RD}$ signal goes low in state T_2, enables the selected memory device, and, after a period of time equal to the access time of the memory, data appears on the data bus. During state T_3 this data is loaded into the instruction register and after this event has taken place $\overline{RD}$ goes high, disabling the memory device. In state T_4 the

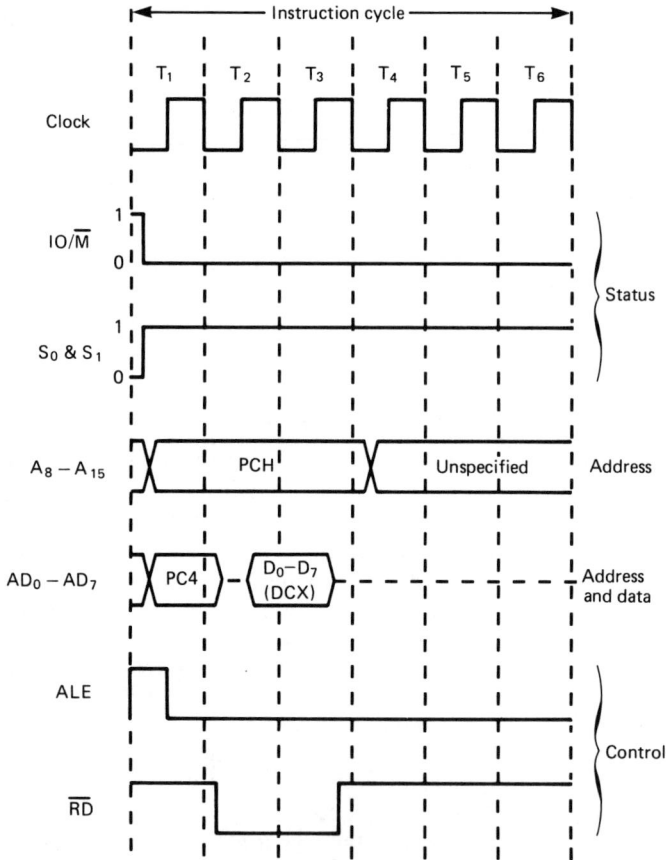

Figure 4.16 Timing diagram for the DCX instruction

microprocessor decodes the opcode previously entered into the instruction register and because the instruction is in this case DCX rp the machine enters state T_4 followed by T_5, thus allowing time for the selected register pair to be decremented.

4.12 The WAIT state

The timing specification of the 8085A states that, when reading from memory or an input device, valid data must appear on the data bus within the time t_r defined by the following equation:

$$t_r = \left[\left(\frac{5}{2} \right) T - 225 \right] \quad \text{ns}$$

and for a 3 MHz clock frequency the periodic time $T = 330$ ns and $t_r = 600$ ns, hence the access time of the memory chip or the I/O device must be $\leqslant 600$ ns.

In practice some devices do not have such short access times; the shorter the access time the more expensive the device will be. In order to accommodate devices with access times >600 ns, a WAIT state T_W is available in the 8085A, entry into this state being controlled by the READY signal.

The state diagram incorporating T_W is shown in Figure 4.17 and it will be observed that if, when in state T_2, READY $= 0$, then on receipt of the next clock pulse a transition will be made to state T_W. The machine now remains in this state until the ready signal goes high, when on the receipt of the next clock pulse it returns to state T_3 and then continues through the machine cycle in the way described earlier. Entry into T_W can be controlled by memory or an I/O device if they generate a ready signal. On entering state T_W, the microprocessor signals existing at the end of state T_2 are frozen for as long as the machine remains in the WAIT state.

The effect of a machine entering T_W is to increase the available access time by discrete amounts governed by the number of clock periods that are spent in the WAIT state. For example, if the machine remains in T_W for one clock period, the available access time becomes 930 ns, for two clock periods it becomes 1260 ns, and so on.

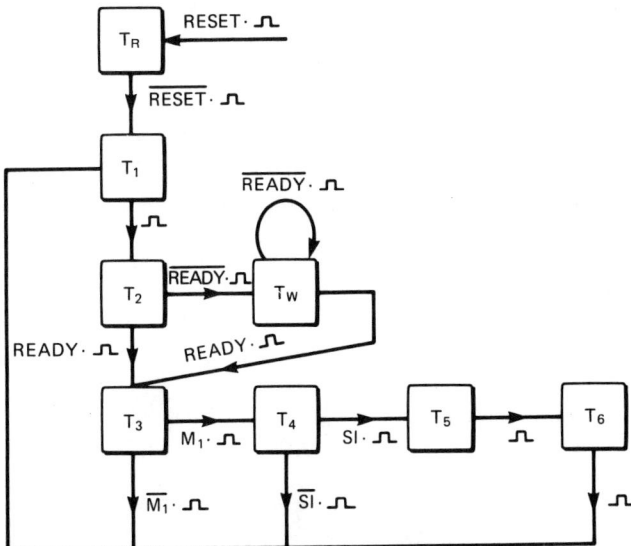

Figure 4.17 State diagram for the 8085A incorporating the WAIT state

A READY signal may also be required during a write cycle. In such a case data is placed on AD_0–AD_7 at the start of state T_2 after the least significant eight address bits have been latched by ALE. The write signal $\overline{WR}$ is lowered at the same time to enable the addressed device. During T_2 the READY signal is sampled to see if a WAIT state is required. If READY = 0, the appropriate number of WAIT states are inserted. On returning to the normal machine cycle in state T_3 the write signal, $\overline{WR}$ goes high, disabling the addressed device and terminating the write operation.

Acceptance of data by the addressed device must take place within a time t_w given by the equation

$$t_w = \left[\left(\frac{5}{2}\right)T - 130\right] \quad \text{ns}$$

after a valid address has appeared on the address pins of the microprocessor. If $T = 330$ ns then $t_w = 695$ ns and for a write time

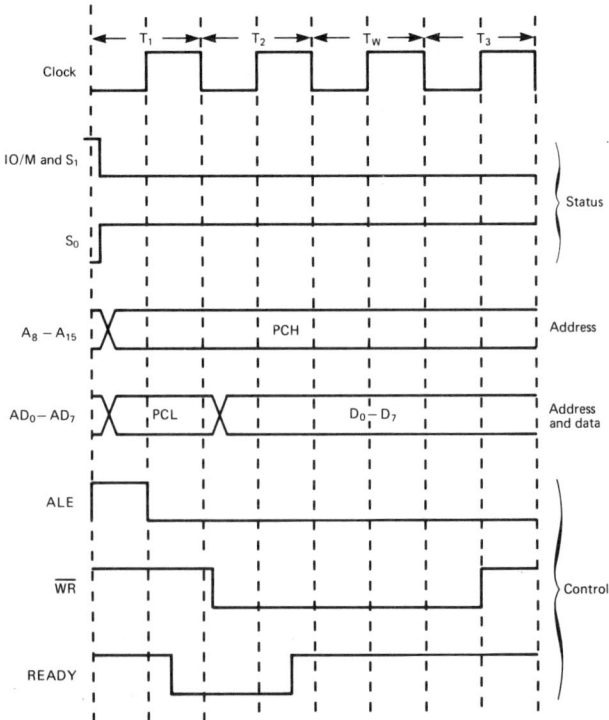

Figure 4.18 Memory write cycle incorporating a WAIT state T_W

longer than this *WAIT* states have to be inserted. A typical example of a memory write machine cycle incorporating a WAIT state T_W is shown in Figure 4.18.

In some cases a READY signal is not required because access time of the selected device is within the previously specified limits of t_r or t_w. When required, the READY signal may be generated by the device that is being accessed or, alternatively, it may be generated by external logic.

4.13 The HALT state

The microprocessor enters the HALT state T_H when the HALT instruction (HLT) is encountered in the program sequence. This is a one-byte instruction consisting of the instruction opcode only, which is transferred to the instruction register in T_3 and is decoded by the instruction decoder in T_4 of the opcode fetch machine cycle. After decoding, the HALT flip-flop within the microprocessor is set and on the next clock pulse the machine returns to state T_1 as illustrated in Figure 4.19. Since in this state HALT = 1 the machine makes a transition to T_H on the next clock pulse, where it remains until the microprocessor is reset or until an interrupt is received.

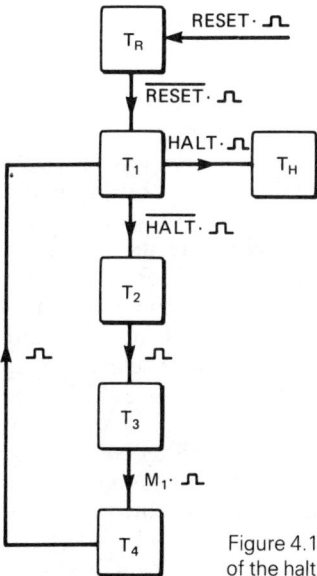

Figure 4.19 State diagram illustrating the implementation of the halt (HLT) instruction

4.14 The SDK 85 expandable microcomputing kit

A typical example of a single-board, expandable, small microcomputing system is the Intel SDK 85. In its basic form the kit consists of:

(a) the 8085A microprocessor,
(b) the 8205, 3-to-8 line decoder,
(c) the 8355, 2 K byte ROM which also incorporates two 8-bit I/O ports that are bit programmable, and
(d) the 8155, 256 byte RWM which incorporates two 8-bit and one 6-bit I/O ports, all of which are programmable. Additionally, this chip contains a 14-bit programmable counter timer.

A block diagram of the basic system is shown in Figure 4.20. Access to memory is via a hexadecimal keyboard which allows a hexadecimal input. Facilities are also provided to display the contents of memory and also the contents of the microprocessor registers.

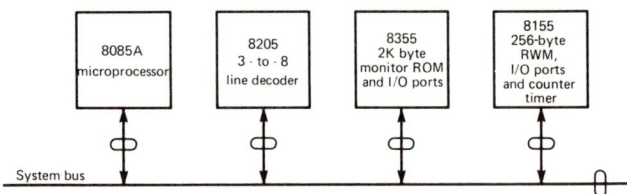

Figure 4.20 Block diagram of the SDK 85

The monitor program in ROM provides what is, in effect, an elementary operating system. It typically contains I/O routines for the hexadecimal keyboard and it is possible to enter, check out and execute small programs. Additionally, it allows breakpoints to be inserted and the microprocessor to be operated in the single-step mode. Facilities are also provided by the monitor to display and modify the contents of memory and machine registers.

The SDK 85 board provides facilities for four memory chips, two 8355s and two 8155s. When one of these four memory chips has to be accessed, then the chip required has to be selected by the appropriate chip enable signals. As it happens, the 8355 has two chip enable pins, of which the one marked $\overline{CE}$ is active low, and the other, marked CE, is kept permanently high by connecting it to the +5 V line, hence chip selection in this case is made via the $\overline{CE}$ pin. In the case of the 8155 there is only one chip enable pin, which is marked $\overline{CE}$ and is active low. The chip to be accessed is identified by means of the 8205, 3-to-8 line decoder and chip select signals are generated by this device in accordance with the tabulation shown in Figure 4.21(a). A basic connection diagram showing the generation and distribution of the chip enable signals in the SDK 85 is shown in Figure 4.21(b).

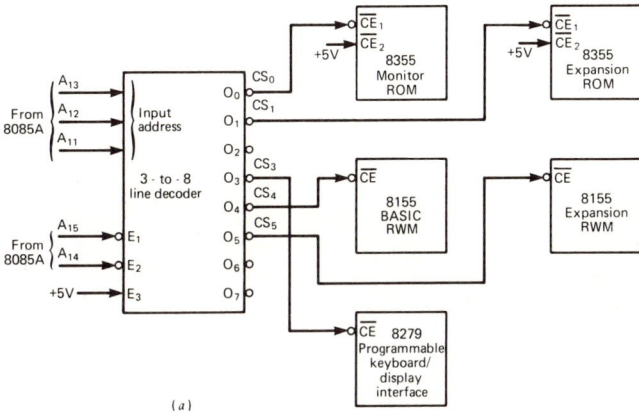

(a)

Output of 3 - to - 8 line decoder	Selected device
CS_0	8355 Monitor ROM
CS_1	8355 Expansion ROM
CS_2	Not connected
CS_3	8279 Keyboard/display controller
CS_4	8155 Basic RWM
CS_5	8155 Expansion RWM
CS_6	Not connected
CS_7	Not connected

(b)

Figure 4.21 (a) The chip enable signals for the SDK 85 (b) Generation and distribution of the chip enable signals

The two most significant address bits A_{15} and A_{14} are connected to the enable lines E_1 and E_2 of the decoder and are maintained permanently low. A_{13}, A_{12} and A_{11} are connected to the address inputs of the decoder, which generates the chip select signals in accordance with the table in Figure 4.21(a). The remaining address bits A_0–A_{10} are used for addressing the memory locations on the individual memory chips. In the case of the 8155, eight address lines are required for the 256 memory locations while for the 8355 11 address lines are required to address the 2048 locations. A connection diagram for a six-chip microcomputer, excluding the keyboard/display controller, is shown in Figure 4.22.

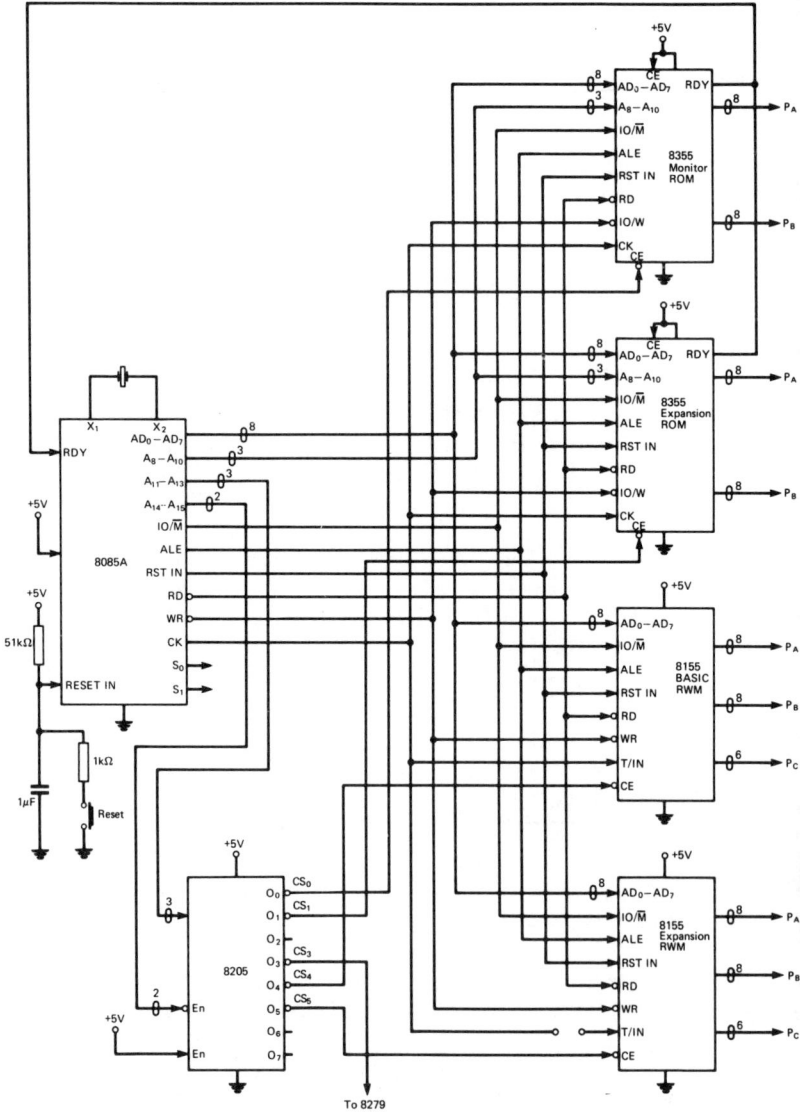

Figure 4.22 Connection diagram for the SDK 85 kit

4.15 Other microprocessors

There are many 8-bit microprocessors presently available on the market, and the following sections will examine briefly the architecture and the principal features of three of them – the Intel 8080A, the Zilog 80 and the Motorola 6800. It will be observed by the reader that there are many common features in all these machines.

4.16 The Intel 8080A

The forerunner of the Intel 8085A was the Intel 8080A. In effect, the 8085A is a single-chip version of the three-chip combination consisting of the 8080A microprocessor, the 8224 clock driver and the 8228 system controller. The 8080A is fabricated in a 40-pin package on which 16 pins are reserved for the address output, and a further eight pins are reserved for connection to the bidirectional data bus. Four pins are required for the $+5$, -5 and 12 volt power supplies and the remaining 12 pins are reserved for timing and control. A pin configuration diagram is shown in Figure 4.23.

Unlike the 8085A, the Intel 8080A employs a two-phase clock as illustrated in Figure 4.24. The time interval between two successive positive-going transitions of phase Φ_1 is defined as the

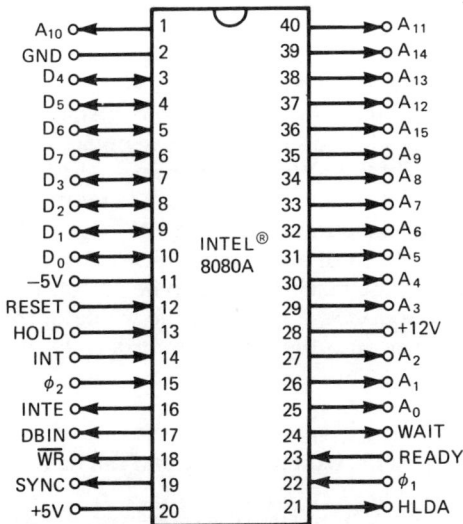

A_{10}	1	40	A_{11}
GND	2	39	A_{14}
D_4	3	38	A_{13}
D_5	4	37	A_{12}
D_6	5	36	A_{15}
D_7	6	35	A_9
D_3	7	34	A_8
D_2	8	33	A_7
D_1	9	32	A_6
D_0	10	31	A_5
$-5V$	11	30	A_4
RESET	12	29	A_3
HOLD	13	28	$+12V$
INT	14	27	A_2
ϕ_2	15	26	A_1
INTE	16	25	A_0
DBIN	17	24	WAIT
$\overline{WR}$	18	23	READY
SYNC	19	22	ϕ_1
$+5V$	20	21	HLDA

(INTEL® 8080A)

Figure 4.23 Pin connections for the 8080A

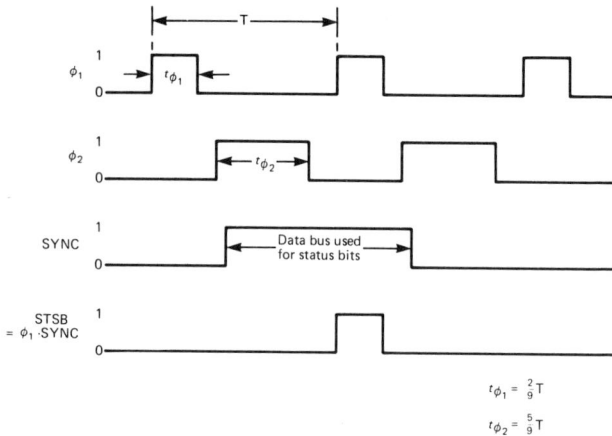

$t_{\phi_1} = \frac{2}{9}T$

$t_{\phi_2} = \frac{5}{9}T$

Figure 4.24 Generation of the status strobe

state time T and it is this clock pulse that divides the machine cycles into states. The clock frequency is nominally 2 MHz and the 8224 clock driver requires an 18 MHz crystal to produce the required clock frequency. Instruction cycle time can now be calculated as for the 8085A. For example, the LDA instruction requires 13 states for implementation, each of time duration 0.5 μs, hence the instruction cycle time is

$$t_i = 13 \times 0.5 = 6.5 \, \mu s$$

a somewhat slower time than for the 8085A.

Timing and logic circuits in the 8080A use the clock signal to generate a SYNC pulse which identifies the beginning of a machine cycle. The transition of the SYNC signal from 0→1 occurs shortly after the 0→1 transition of the Φ_2 clock signal and it lasts for a period of T seconds. For the duration of the SYNC signal the data bus is used to carry a pattern of status bits which inform the outside world of the nature of the machine cycle to be performed. Since the status bits may be required throughout the whole of the machine cycle they are latched into the system controller by the *status strobe* (STSB) where

$$STSB = \Phi_1.SYNC$$

Once the status bits are latched the data bus is free and can be used as a highway for data.

Like the 8085A, the 8080A has six general purpose 8-bit registers – B, C, D, E, H, and L – a program counter, a stack pointer and a status or flat register. This group of registers is the one that can be referred to explicitly when writing a program. The general purpose registers can be used as either 8-bit registers or 16-bit register pairs. Additionally, there is an instruction register, an 8-bit temporary accumulator which provides temporary storage for the second operand before input to the ALU, and also temporary registers W and Z performing the same function as they do in the 8085A. Since the address and data bus are not multiplexed, separate address and data buffers are provided.

The functional block diagram of the 8080A microprocessor is shown in Figure 4.25 and it will be seen that it is not dramatically different from the corresponding diagram for the 8085A illustrated in Figure 4.12. The major differences appear in the timing and control section where two signals DBIN and $\overline{WR}$ are generated. These signals are taken to the 8228 system controller as shown in Figure 4.26, which illustrates the standard CPU interface for this machine. DBIN indicates to external circuits that the bidirectional data bus is in the input mode, while $\overline{WR}$ indicates that the 8080A has placed data on the data bus and that it is stable. Additionally, the SYNC signal is generated the function of which has been described earlier.

The 8228 system controller generates the memory and I/O signals that are required, latches the status bits, and acts as a bidirectional bus driver, as indicated in the schematic diagram shown in Figure 4.27. The memory and I/O signals generated by this chip are: memory read $\overline{MEMR}$, memory write $\overline{MEMW}$, I/O read $\overline{I/OR}$ and I/O write $\overline{I/OW}$, which control the transfer of data in both directions between memory and microprocessor and I/O and microprocessor. These signals are identical to those which can be generated in the 8085A by logically combining the IO/$\overline{M}$ status signal and the $\overline{RD}$ and $\overline{WR}$ control signals of that machine, as described earlier in this chapter.

There are 72 distinct instructions in the instruction set of the 8080A, two less than the 74 provided with the 8085A. The 8085A has more comprehensive interrupt facilities and the two additional instructions are concerned with the setting and reading of the interrupt masks and the serial input/output features of the machine. In spite of the differences in the instruction sets, the two machines are software compatible, and any program developed for the 8080A can be run on the 8085A. However, it should be noted that, assuming both machines operated at the same clock frequency, program execution times would be different because implementation of identical instructions on the two machines requires, in some cases, a different number of states.

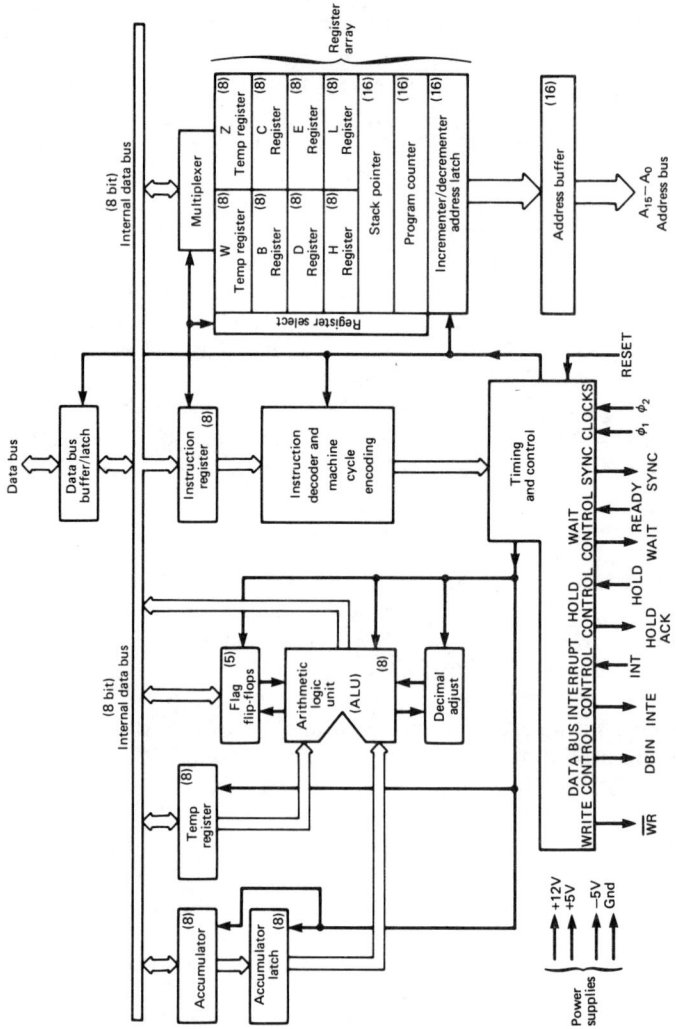

Figure 4.25 Functional block diagram of the 8080A

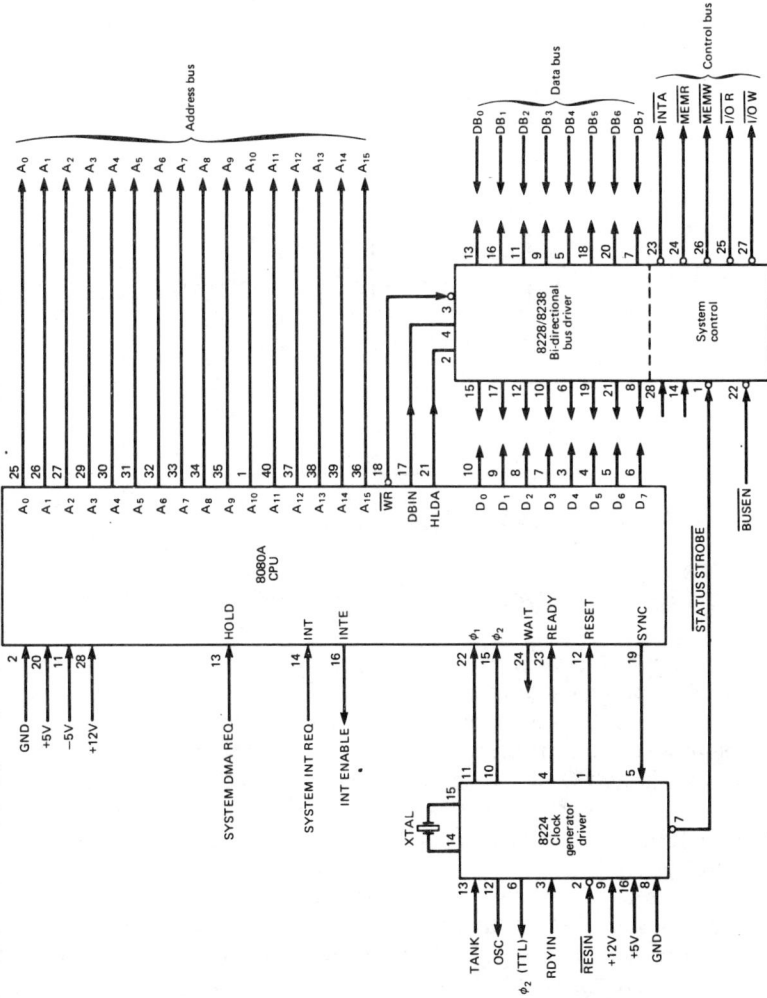

Figure 4.26 CPU standard interface

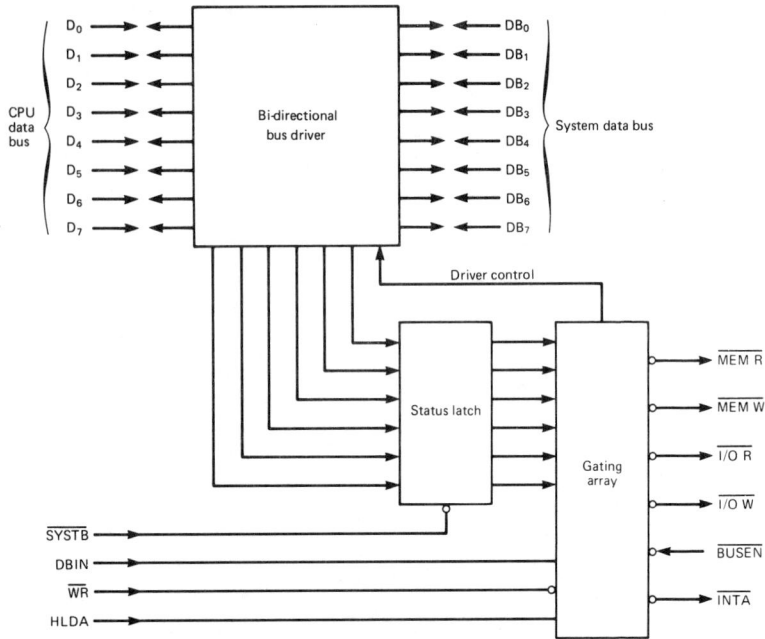

Figure 4.27 Block schematic of the 8228 system controller

4.17 The Zilog Z80

The Zilog Z80, also an 8-bit microprocessor, provides many other facilities not available in the 8080A/8085A. However, the Z80 is, by and large, software compatible with these two machines so that programs written for them can be run on the Z80. Compared with the 8080A/8085A, the Z80 has a greatly enlarged instruction set consisting of 158 instructions in all. Many of the additional instructions are logical extensions to the 8080A/8085A instruction set but many are completely new in concept. The Z80 duplicates the eight registers provided in the 8080A/8085A although only one set of registers can be operational at any one time. Additionally, there are two index registers for indexed addressing, an interrupt vector register providing special interrupt facilities, and a memory refresh register giving a dynamic memory refresh capability. The provision of new instructions, new addressing modes and new hardware features results in a much more versatile machine.

There are 14 general purpose 8-bit registers associated with the Z80 – A, B, C, D, E, H and L, and A′, B′, C′, D′, E′, H′ and L′ –

and there are also two flag or status registers, F and F'. Special instructions are available for selecting between A and F and A' and F', and for selecting between the remaining two groups of registers. Such an arrangement has advantages not available in the two previously described machines. The machine programmer may switch from one bank to the other and make more register storage available. This eliminates the unnecessary transfer of temporary data from CPU to memory and back, resulting in a decrease in program execution time. Additionally, the alternative bank of registers can be used for storing the status of the machine during an interrupt service routine.

The 8-bit accumulator has the same function as in the 8080A/8085A and the general purpose registers can be operated singly or as a pair which can, if necessary, act as a pointer to data stored in memory. A status or flag register provides the decision-making information for the machine. For the Z80 the parity flag has the dual function of specifying both parity and overflow and there is also a subtract flag which is set during BCD subtract operations.

Specialized registers common to all three machines are available in the Z80. These are the instruction register, the program counter and the stack pointer. Two 16-bit index registers are provided for the indexed addressing mode and a *memory refresh register* is available, which allows automatic refreshing of dynamic memory,

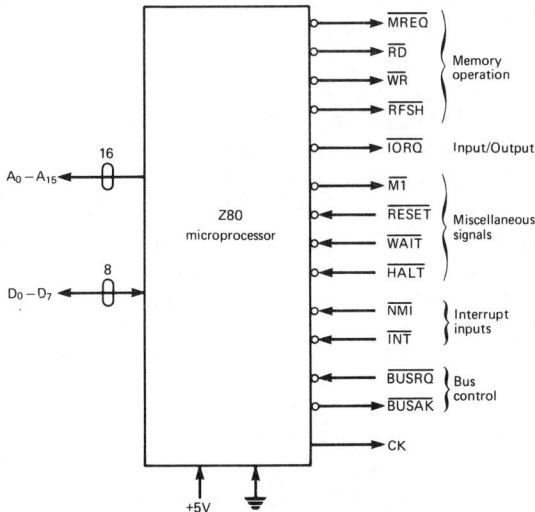

Figure 4.28 Z80 microprocessor pin connections

when used, every 2 ms. Special interrupt facilities are provided by the *interrupt vector register*. This register specifies eight bits which, when combined with a further eight bits supplied by the interrupting device, provides the starting address of an interrupt service routine.

Figure 4.28 illustrates the various connections to the Z80, 40-pin package. Address and data buses are not multiplexed so that separate pins are provided for address and data signals. Additionally, there are 13 control signals associated with the CPU which can be divided into a number of distinct groups.

The signals associated with memory operation are, first, $\overline{\text{MREQ}}$, an active low signal which indicates that a valid memory address is appearing on the address bus. It is, in effect, part of the chip enable signal requesting memory either to output data for a memory read or to input data for a memory write. Control signals $\overline{\text{RD}}$ and $\overline{\text{WR}}$, both active low, indicate to memory whether the operation to be implemented is either a read or write. Finally, in this group of signals $\overline{\text{RFSH}}$ is used when dynamic memories are employed to implement the periodic refreshing of stored data.

There is a single I/O signal $\overline{\text{IORQ}}$ which is used for input/output requests. When this signal is low then $\overline{\text{RD}}$ and $\overline{\text{WR}}$ are used to determine whether the I/O operation is to be an I/O read or write.

A status signal $\overline{\text{M1}}$ is provided and indicates that the microprocessor is performing the first machine cycle of an instruction which is, of course, the opcode fetch. Other miscellaneous signals available are $\overline{\text{RESET}}$, which is used as the CPU master reset, $\overline{\text{WAIT}}$, which allows the insertion of wait states for slow memories, and $\overline{\text{HALT}}$, which goes low when the halt instruction is encountered in the program sequence.

Two signals are available for interrupt operations. The first is $\overline{\text{NMI}}$ which cannot be disabled and is always recognized by the CPU at the end of the current instruction. There are exceptions to this statement, for example if a $\overline{\text{BUSRQ}}$ signal has been received, or, alternatively, if the machine is in a continuous wait state. The main interrupt request signal $\overline{\text{INT}}$ will be recognized at the end of the current instruction if the interrupt enable flip-flop has been set and if $\overline{\text{BUSRQ}}$ is not active.

The remaining two signals are concerned with bus control. $\overline{\text{BUSRQ}}$ is generated by an external device which requires to take over control of the bus system. A typical example of when such a requirement is needed occurs during a DMA operation where the DMA controller acts as a second microprocessor and takes over control of the system bus. The $\overline{\text{BUSAK}}$ signal is the response to $\overline{\text{BUSRQ}}$ and indicates that the data, address and control outputs of the Z80 are in the high impedance state and the system bus can now be controlled by the external device.

4.18 The Motorola 6800

The 6800 processor is manufactured on a single chip using N-channel, silicon gate MOS technology. The main features of the device are that it processes eight bits of data in parallel, has a 16-bit address bus and can access a maximum of 64K bytes of memory. The instruction set contains 72 different instructions and four different addressing modes are available, including indexed addressing.

It should be observed that there are no general purpose registers in this machine as there are in the 8080A, the 8085A and the Z80. However, the processor does contain two accumulators, A and B, which are used for all arithmetic and logical operations. Specialist registers such as the instruction register, program counter and stack pointer are available, and a 16-bit index register is provided which is used in the indexed addressing mode. A condition code register CCR stores information describing various aspects of the results previously generated in the ALU. This register contains three extra flags in addition to the normal ones, namely, carry, zero and sign. These are O, a 2's complement overflow flag, which is set when a carry occurs from bit 6 to bit 7 in the ALU, H, the half-carry flag used for adjusting the result of BCD addition, and I, the interrupt mask flag which indicates whether interrupts have been enabled.

In addition to the registers described above there are three others which, like the instruction register, are not accessible to the programmer. They are the *data address register*, which is used for temporary storage of the address of an operand specified by an instruction, and also an 8-bit temporary register associated with the ALU, which holds one of the operands prior to and during arithmetic or logical operations.

The processor requires a two-phase clock which governs the instruction execution time, time available for memory access, and the time at which data transfers occur on the data bus. To function correctly, voltage levels, frequency, width, overlap and rise time of the two clock signals Φ_1 and Φ_2 must be within closely specified limits. For all practical purposes these two signals can be regarded as being exactly out of phase.

Figure 4.29 illustrates the pin connections for the 6800. There are 16 address pins, eight data pins and nine control pins. Two pins are reserved for the two phases of the clock and three pins are reserved for the 5 V power supply, leaving two spare pins.

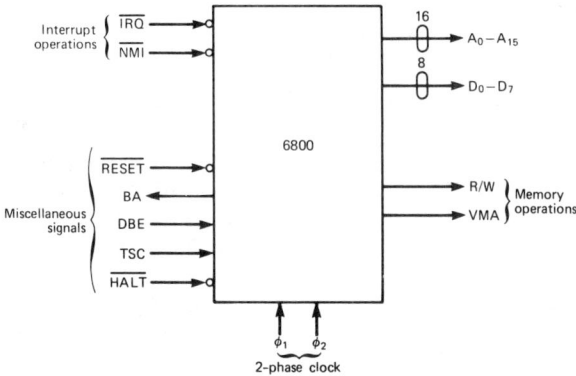

Figure 4.29 6800 microprocessor pin connections

The valid memory address signal VMA is active during each clock cycle involving a bus transfer; when this signal is high it indicates that the address on the bus is valid, while if it is low the 6800 is free to perform internal operations. A second signal R/W indicates the direction of data transfer. If R/W = 1 a read operation is being performed, while if R/W = 0 the processor is writing to an external device. During a read operation the processor data bus drivers are automatically disabled internally so that another device such as memory can place data on the bus. The processor, however, will only do so if the data bus enable (DBE) control signal is high. Since all data transfers occur during the second half of a clock cycle, DBE is normally connected to phase Φ_2 of the clock. A timing diagram for a write cycle when DBE is connected to Φ_2 is shown in Figure 4.30.

The $\overline{\text{RESET}}$ input provides a means of starting the program counter, and hence the program, from a predefined address. When the $\overline{\text{HALT}}$ signal is low the processor terminates its activities at the end of the current instruction and the bus system could then, if required, be available for a DMA operation. The output bus available (BA) signal indicates when the processor has stopped, and it can be kept in this state indefinitely without a loss of data while the DMA operation is being performed.

The tri-state control (TSC) signal is also used for DMA

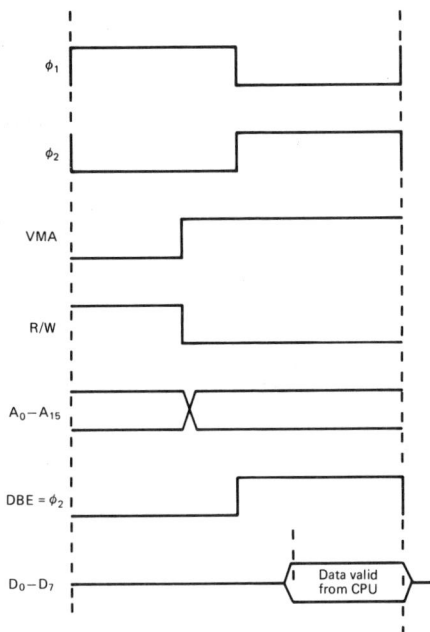

Figure 4.30 The Write cycle for the 6800 when DBE = Φ_2

operations. This signal is set high on the rising edge of Φ_1 and places the processor address bus outputs in the high impedance state. If DBE is connected to Φ_2 then, when Φ_1 goes high, the processor data outputs are also in the high impedance state and consequently both data and address bus are available for a DMA transfer. However, the processor can only be maintained in this state for a maximum time period of 4.5 µs, otherwise data stored in dynamic memory will be lost.

The remaining two control signals $\overline{IRQ}$ and $\overline{NMI}$ are associated with interrupt operations. Normally, interrupts are initiated via the $\overline{IRQ}$ input, which responds to a logic O level if the interrupt mask flag is not set. The $\overline{NMI}$ interrupt input has an almost identical effect but the processor responds to a logic O at the input regardless of the state of the interrupt flag.

Problems

4.1   Microprocessors are frequently referred to as 8-bit or 16-bit machines. Explain the significance of this statement.

4.2   Discuss the differences between ROM and RWM and mention the functions these devices perform in a microprocessor system.

4.3   The address bus in a microprocessor system is normally unidirectional while the data bus is bidirectional. Give reasons for this and discuss the advantages of the bus system.

4.4   Write a brief description of the function of the following 8085A registers: (*a*) program counter, (*b*) instruction register, (*c*) accumulator, (*d*) stack pointer and (*e*) scratchpad register.

4.5   Draw a block diagram of an 8-bit register having a tri-state output for each stage of the register. Show how a group of four such registers can be connected to a common bus register, selection being controlled by a decoder.

5 The instruction set I: data transfer, arithmetic and logic operations

5.1 Introduction

Each microprocessor has its own instruction set specified by the manufacturer. Instruction sets vary from machine to machine, but there are many instructions which are common to all machines. It is common practice to classify instructions into a variety of groups which emphasize specific features of the instruction set, such as data transfer or data manipulation. One of the functions of this chapter will be to investigate the instruction groupings and examine the implementation of particular instructions associated with the various groups.

An instruction consists of two *fields* as indicated below:

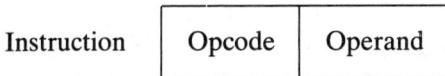

Instruction	Opcode	Operand

They are the opcode field and the operand field. The opcode field identifies the operation to be performed by the machine, while the operand field indicates by a variety of methods the address of the data on which the operation is to be performed. The methods for specifying the address of the operand are referred to as addressing modes. Most machines provide a number of addressing modes which are designed to give the programmer maximum flexibility when writing programs, and a second aim of this chapter will be to present in some detail the addressing methods commonly in use and, in particular, those methods available in the 8085A.

5.2 Machine code representation of instructions

The instruction is a fundamental element of a program listing and it must specify two things: (a) the operation to be performed; and (b) the operand.

In its most basic form it is represented by a pattern of binary digits which, in the case of the 8085A, may be one, two or three bytes long. This form of instruction representation is called machine code and a typical example of a short program written in machine code is given below:

$$
\left.
\begin{array}{ll}
0011 & 1010 \\
0110 & 1000 \\
0000 & 0000
\end{array}
\right\} \text{Instruction 1}
$$

$$
\begin{array}{ll}
0010 & 1111
\end{array} \quad \text{Instruction 2}
$$

$$
\left.
\begin{array}{ll}
0011 & 0010 \\
0110 & 1001 \\
0000 & 0000
\end{array}
\right\} \text{Instruction 3}
$$

$$
\begin{array}{ll}
0111 & 0110
\end{array} \quad \text{Instruction 4}
$$

In this instance the program is written in 8085A machine code; its function is to complement the contents of a specified memory location and, after complementation, to place the results in a second specified memory location. The purpose of this program would only be obvious to the programmer, and even then only for a limited time.

Although the machine operates with the above 8-bit patterns there are serious disadvantages to this type of representation. Programming in machine code for any but the shortest programs is very tedious, and furthermore it is prone to errors which are not easily recognizable. Additionally, loading a program into memory in binary form is also tedious and prone to error since it would have to be done via eight binary switches.

The short piece of program given previously has been separated into individual instructions and it will be observed that two of them are three bytes long and two are one byte long. The first byte of instruction 1 identifies the operation, in this case 'Load the Accumulator', while bytes 2 and 3 identify the operand which is specified by the address given in these two bytes.

Some of the disadvantages of machine code programming can be avoided by writing the program in hexadecimal notation. For example, a more concise form of the machine code program in hexadecimal form is given below:

```
3A   68   00   Instruction 1
2F              Instruction 2
32   69   00   Instruction 3
76              Instruction 4
```

Writing programs in this form is less tedious and not so prone to error. Furthermore, programmers, after a little experience of the instruction set, are readily able to recognize individual instruction opcodes. For example, the single-byte instruction 76 will soon be recognized as the 8085A instruction for Halt, and, similarly, 3A represents the instruction 'Load the Accumulator', while 0068H specifies the address of the data to be loaded.

An additional reason for using hexadecimal notation is that some machines such as the SDK 85 have hexadecimal keyboards, so that after writing a program directly in hexadecimal form it can be entered into memory using the keyboard. If this facility has been provided, the translation from hexadecimal to binary has to be accomplished within the machine and this is done in the SDK 85 under the control of the monitor program.

5.3 Mnemonic representation of instructions

The mnemonic form of an instruction can, after the hexadecimal representation, be regarded as its next higher level of representation. A typical example of the mnemonic form is LDA, which can be interpreted as 'Load the accumulator from memory'. Instruction cards supplied by the manufacturer will give the mnemonic form of the instruction and its corresponding hexadecimal representation, whereas, for the 8085A, the instruction set gives the mnemonic and binary representations of the instruction.

Without, at this stage, going into the detailed rules for writing programs in mnemonic form, the complementation program previously written in machine code and hexadecimal form would appear in mnemonic form as shown below:

```
LDA   68H   Instruction 1
CMA         Instruction 2
STA   69H   Instruction 3
HLT         Instruction 4
```

This type of program listing is referred to as an *assembly language* program, or, alternatively, as a program written in *symbolic language*.

Writing an assembly language program is clearly much easier because the mnemonic representation bears a close relationship to the instruction it represents. For example, STA simply means 'Store the contents of the accumulator in a specified memory location'. Because of this relationship the programmer will find it easier to commit mnemonic forms to memory rather than a string of binary digits or a pair of hexadecimal digits.

Before an assembly language program can be executed it has to be translated into machine code. After modification and correction, the programmer can do this with a conversion table, the process being referred to as *manual assembly*. However, it is much more likely that the translation will be performed by the machine itself with the aid of an assembler program.

Besides the binary and mnemonic representations of the instruction, a great deal more information is given by the instruction set. For example, the entry under the 'Add Register' heading is given in full below:

ADDr   (Add Register)
$(A) \leftarrow (A) + (r)$
The content of register r is added to the content of the accumulator. The result is placed in the accumulator.

1	0	0	0	0	S	S	S

Cycles:	1
States:	4
Addressing:	Register
Flags:	Z, S, P, CY, AC

The second line of this entry is a register transfer language statement describing an arithmetic micro-operation. The right-hand side of the equation is interpreted as the contents of the accumulator (A) added to the contents of register (r), where the brackets indicate 'the contents of'. The arrowhead pointing left is called the *replacement operator* and is indicating that the result of the arithmetic micro-operation is to become the contents of the accumulator.

DDD or SSS	Register
111	A
000	B
001	C
010	D
011	E
100	H
101	L

Figure 5.1 Coding of destination or source registers in the 8085A

The binary representation of the instruction is shown inserted in a conventional register representation and it will be noticed that the entry in the three least significant places of the register is S. This letter S indicates the source register and the appropriate code for this register is obtained from the table shown in Figure 5.1.

If, for example, the contents of register B are to be added to the contents of the accumulator A the binary representation of the instruction will be

1	0	0	0	0	0	0	0

and the corresponding hexadecimal representation is

8	0

It is worth noting that for certain instructions such as MOV r_1r_2 which, when implemented, results in the contents of register r_2 being moved into register r_1, the binary representation given in the instruction set is

0	1	D	D	D	S	S	S

where $r_1 = D$, the destination register, and $r_2 = S$ is the source register. If the requirement is to move the contents of the accumulator A to register B then the binary configuration becomes

0	1	0	0	0	1	1	1

and the corresponding hexadecimal representation is

4	7

The rest of the information given regarding the ADD instruction indicates that it is executed in one machine cycle consisting of four T-states. Finally, the last line of the instruction, headed 'Flags', indicates that after the add operation has been performed in the ALU all the flags will be modified in accordance with the result placed in the accumulator.

If the contents of register B are F2H and the contents of the accumulator are ABH then the operation performed by the machine is illustrated below:

```
F2   1 1 1 1 0 0 1 0   Reg. B
AB   1 0 1 0 1 0 1 1   Reg. A
───────────────────
9D   1 0 0 1 1 1 0 1   Reg. A
```

and the condition of the flats after the operation will be

Carry = 1;	CY set to 1
Sign bit (MSB) = 1;	S set to 1
Non-zero answer;	Z reset to 0
Result in register A has odd parity;	P reset to 0
No carry from bit 3 to bit 4;	AC reset to 0

There are some instructions which involve register pairs such as INX rp (increment register pair). The binary configuration given for this instruction in the instruction set is

0	0	R	P	0	0	1	1

and the coding for register pairs in the 8085A is given in the table shown in Figure 5.2.

RP	Register pair
00	BC
01	DE
10	HL
11	SP

Figure 5.2 Coding for register pairs in the 8085A

If the above instruction had been related to the register pair HL, then the binary and hexadecimal representations of the instruction would be

0	0	1	0	0	0	1	1	=	2	3

5.4 Addressing modes

Most microprocessors provide a number of different addressing modes. The 8085A, for example, has the following addressing modes available:

(1) Implied addressing
(2) Immediate addressing
(3) Direct addressing
(4) Register and register indirect addressing
(5) Modified page zero addressing.

Common addressing modes employed in other machines are:

(1) Indexed addressing
(2) Relative addressing.

Although the above two addressing modes are not directly available in the 8085A, it is possible, for example, to create indexed addressing by combining more than one instruction.

The function of an addressing mode is to provide the address of an operand. Some instructions only require one operand and its location is known since it is either in the accumulator or one of the general purpose registers. However, many instructions require two operands, and, while the address of one is known, the address of the second one has to be specified by the instruction. The method of specifying this address is referred to as the addressing mode, and those modes previously listed will now be examined in some detail.

5.5 Implied addressing

This is not an addressing mode in the real sense of the word since neither an operand or its address have to be specified by the instruction. It normally refers to a single-byte instruction which simply defines an operation that has to be performed on the contents of the accumulator or one of the general purpose registers. Typical examples of this type of instruction are complement, increment and decrement. For example, CMA, the

Byte 1 | 0 | 0 | 1 | 0 | 1 | 1 | 1 | 1 | = | 2 | F | Opcode

Figure 5.3 Immediate addressing in the 8085A, the CMA instruction

complement accumulator instruction in the 8085A, the format of which is shown in Figure 5.3, directs the contents of the accumulator to the ALU, inverts them, and returns them to the accumulator.

5.6 Immediate addressing

In this type of addressing, the opcode and the operand are stored in consecutive memory locations. The opcode occupies the first byte of the instruction and the operand occupies the second byte. Instructions using this mode of addressing are clearly executed faster than those which have to bring the address of an operand from memory.

A typical example of an 8085A instruction that uses the immediate addressing mode is MVI, the move immediate

instruction. The instruction consist of two bytes; the opcode field
in the first byte is variable, and consequently the data in the
second byte can be moved to the accumulator or any of the general
purpose registers. The format of the instruction is shown in Figure
5.4(*a*) where D represents the destination register. If the data is to
be moved to the accumulator then the format of the instruction
would be as shown in Figure 5.4(*b*).

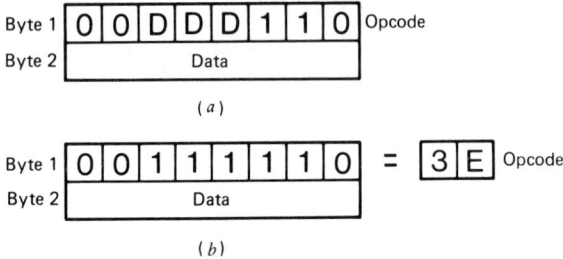

Byte 1 | 0 | 0 | D | D | D | 1 | 1 | 0 | Opcode
Byte 2 | Data

(*a*)

Byte 1 | 0 | 0 | 1 | 1 | 1 | 1 | 1 | 0 | = | 3 | E | Opcode
Byte 2 | Data

(*b*)

Figure 5.4 Immediate addressing: (*a*) Instruction format (*b*) Transfer of data to
accumulator

An *extended* form of immediate addressing is also available in
the 8085A. In this case the instruction consists of three bytes,
where bytes 2 and 3 contain the data that has to be moved to a
register pair specified by the opcode. A typical example of
extended immediate addressing in the 8085A is provided by the
instruction LXI, load register pair immediate. If the data is to be
loaded into the 16-bit stack pointer then the instruction format is
as shown in Figure 5.5.

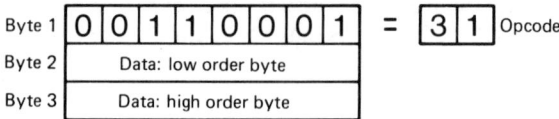

Byte 1 | 0 | 0 | 1 | 1 | 0 | 0 | 0 | 1 | = | 3 | 1 | Opcode
Byte 2 | Data: low order byte
Byte 3 | Data: high order byte

Figure 5.5 Direct immediate addressing: load data into stack pointer

5.7 Direct addressing

In the direct addressing mode, the address of the operand is
contained in the instruction itself. Since the address normally
consists of 16 bits, two bytes are required to hold it, and hence a
three-byte instruction sequence is required. A typical example of
the direct addressing mode is provided by the 8085A instruction

LDA, load the accumulator. The first byte of the instruction is the opcode, and the second and third bytes of the instruction contain the address of the data to be loaded into the accumulator. The instruction format is illustrated in Figure 5.6 and it will be observed that byte 2 of the instruction contains the least significant eight bits of the address and byte 3 contains the most significant eight bits.

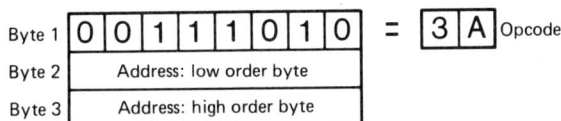

Byte 1 | 0 | 0 | 1 | 1 | 1 | 0 | 1 | 0 | = | 3 | A | Opcode

Byte 2   Address: low order byte

Byte 3   Address: high order byte

Figure 5.6 Direct addressing: load accumulator instruction

Another example of direct addressing in the 8085A is the LHLD instruction. The fuction of this instruction is to load the contents of the memory location whose address is specified by bytes 2 and 3 of the instruction into register L. The contents of the memory location at the succeeding address is loaded into register H. The whole operation is described by the two following register transfer language statements

$$L \leftarrow ((\text{byte } 3)(\text{byte } 2))$$
$$H \leftarrow ((\text{byte } 3)(\text{byte } 2) + 1)$$

Page-zero direct addressing mode is similar to direct addressing, except that it requires a two-byte instruction and the least significant eight bits of the address containing the operand are contained in byte 2 of the instruction. It is then assumed that the eight most significant bits of the address are 00H. Using this method it is clear that the operand can only be stored within the address range 00–FFH, that is, the first $(256)_{10}$ memory locations.

5.8 Register and register indirect addressing

For the register addressing mode one or more of the microprocessor registers is addressed by the instruction. The instruction opcode contains a three-bit field that specifies a general purpose register, and the operand is located in that register. A typical example of the use of this addressing mode in the 8085A is the ANA instruction, by which the contents of a specified register are ANDed with the contents of the accumulator. It is a single-byte instruction, as illustrated in the instruction format shown in Figure 5.7(a) where SSS represents the address of the source register for the second operand. If the source of the second operand is register H, then the instruction format is completely specified by the replacement of SSS by 100, as shown in Figure 5.7(b). When implemented, the instruction will logically AND the contents of register H with the contents of the accumulator and place the result in the accumulator.

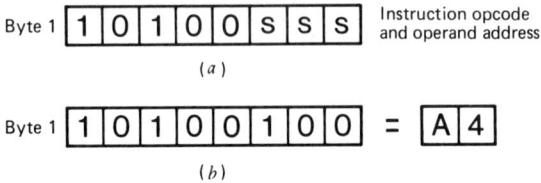

Byte 1 | 1 | 0 | 1 | 0 | 0 | S | S | S | Instruction opcode and operand address

(a)

Byte 1 | 1 | 0 | 1 | 0 | 0 | 1 | 0 | 0 | = | A | 4 |

(b)

Figure 5.7 Register addressing: (a) Format of ANA instruction (b) AND contents of H register with contents of accumulator

In the case of the register indirect addressing mode, a register pair is employed as a pointer to memory. For the 8085A any of the register pairs BC, DE or HL may be used for this purpose. This kind of addressing is extremely useful when the data is grouped in a block of consecutive memory locations. The method is not efficient if the data is not in a block, since the pointer register must then be loaded with the address of every byte of data that has to be accessed.

An example of this type of addressing is provided in the 8085A by the LDAX, load accumulator indirect instruction. The content of the memory location, whose address is in the register pair specified by the instruction opcode, is moved to the accumulator. If the HL registers are the ones specified by this instruction, then the format of the instruction is as illustrated in Figure 5.8. It will be observed that LDAX is a single-byte instruction which contains a two-bit field 10 in bits 4 and 5, that identify the register pair, HL, holding the address of the operand.

It is also possible in the 8085A to create a register indirect addressing mode by the combination of a pair of instructions. For

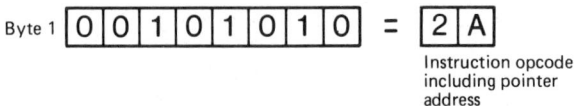

Byte 1 | 0 | 0 | 1 | 0 | 1 | 0 | 1 | 0 | = | 2 | A |

Instruction opcode including pointer address

Figure 5.8 Register indirect addressing: the LDAX instruction

example, the instruction LHLD can be used to load the address of the data into the H and L registers. If it is then followed by MOV, the move instruction, the data in the location pointed at by the H and L registers is transferred to the location specified by the instruction MOV. To transfer the contents of memory location 5348H to the accumulator requires the following assembly language program for an indirect load:

LHLD 5348H ; Load HL with data address
MOV A,M   ; Move data from memory to accumulator
HLT   ; Stop

5.9 Indexed addressing

This mode of addressing is not directly available in the 8085A, although if the programmer considers indexed addressing essential it is possible by combining instructions to create this mode. In this type of addressing the effective or actual data address can be obtained by adding the *address offset* contained in the instruction to the *base address* stored in the index register. For example, if

Address offset   = 29H
Base address   = 0179H
Effective address = Address offset + Base address
                 = 29 + 0179
                 = 01A2

If the instruction is to store the contents of the accumulator, then they will be placed in memory location 01A2H. The format of the two-byte instruction is illustrated in Figure 5.9.

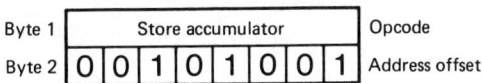

Byte 1 | Store accumulator | Opcode
Byte 2 | 0 0 1 0 1 0 0 1 | Address offset

Figure 5.9 The indexed addressing mode

The indexed addressing mode is used in the 6800 and is implemented in the way described above. The address offset, provided by the second byte of the instruction, is always interpreted as a positive integer. Since eight bits are available for the address offset its value will lie in the range 00–FFH.

In the Z80 the signed 2's complement convention is used to represent the address offset, and it can therefore take on a range of values from $(-128)_{10}$ to $(+127)_{10}$. If, for example, the contents of the index register are 5D01H and the address offset is 0AH then

Effective address = 0A + 5D01
                   = 5D0BH

If, however, the address offset is F0H, then

Effective address = F0 + 5D01
                   = 5C91H

In the second example, the most significant bit of the address offset is 1, hence it is a negative quantity and has to be deducted from the base address.

When an instruction employing the indexed addressing mode is implemented, the addition of the contents of the 16-bit index register to the address offset does not alter the contents of the index register. The microprocessor calculates the effective address and places it on the address bus, but it is lost after the implementation of the instruction.

The indexed addressing mode can be created in the 8085A in the following manner. Registers H and L are employed as a 16-bit index register, while the general purpose registers D and E can be programmed to hold the address offset. The contents of the two registers are then added together using the DAD instruction, thus providing the operand address in the H and L registers. The following assembly language program illustrates the implementation of the indexed addressing mode with the 8085A.

```
LHLD   2000H   ; Load HL with base address
LXI    OAH     ; Load DE with address offset
DAD            ; Add contents of HL and DE for effective
                 address
MOV    A,M     ; Move contents of location addressed by HL
                 to the accumulator
HLT            ; Halt
```

5.10 Relative addressing

In the relative addressing mode the address portion of the instruction is a single byte, thus reducing the instruction length by one byte when a comparison is made with direct addressing, and hence reducing the storage requirement. The method allows the addressing of 256 locations around the current address, see Figure 5.10(a). The instruction format is shown in Figure 5.10(b) and since the number stored in the displacement byte is signed, it has a range from $(-128)_{10}$ to $(+127)_{10}$. Because the instruction consists of two bytes, when the displacement value is added to the contents of the program counter a memory location from $(-126)_{10}$ to $(+129)_{10}$ bytes relative to the current instruction may be obtained.

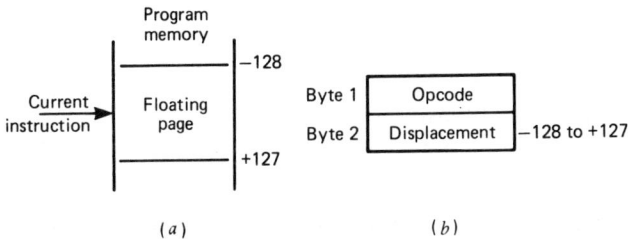

Figure 5.10 Relative addressing: (a) Concept of 'floating page' (b) Instruction format

During program execution the current instruction moves through consecutive memory locations, and the block of memory that can be addressed by the relative addressing technique moves along with the current instruction. The block consists of 256 memory locations and hence constitutes a 'floating page'.

5.11 Modified page zero addressing

Modified page zero addressing is available in both the 8085A and the Z80 microprocessors. It is provided with one instruction only, namely RST, restart. The format of the instruction is shown in

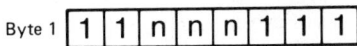

Figure 5.11 Modified page zero addressing instruction format

Figure 5.11 and the effect of the instruction is to cause a jump to a new address on page zero 00–FFH, which is defined by the value of nnn, after pushing the present contents of the program counter onto the stack. There are eight possible combinations of nnn, which allow a jump to locations 00H, 08H, 10H, 18H, 20H, 28H, 30H and 38H.

5.12 Instruction groupings for the 8085A

The instructions associated with the 8085A can be conveniently divided into five basic groups. They are:

(a) Data transfer
(b) Arithmetic
(c) Logic
(d) Branch
(e) Stack and machine control.

These various groups of instructions will now be examined in some detail, beginning with those that are associated with data transfers.

5.13 Data transfer instructions

Data transfer instructions are concerned with three main activities:

(1) Transfers between microprocessor and memory in both directions
(2) Inter-register transfers within the microprocessor
(3) Transfers between microprocessor and its associated peripherals in both directions.

These various types of transfer can be achieved in a variety of ways:

(1) Transfers employing the immediate addressing mode
(2) Data transfers between source and destination registers
(3) Transfers employing the direct addressing mode
(4) Transfers employing the register indirect addressing mode.

5.14 Transfers employing immediate addressing

The objective of load immediate instructions is to transfer a byte of data directly from memory into a microprocessor register. The first byte of the instruction indicates to the microprocessor that the current instruction is an immediate one and it also specifies the register to be loaded, while the second byte contains the data to be loaded.

Figure 5.12(a) shows a block schematic of all the general purpose registers of the 8085A, and additionally the accumulator, the status register, the stack pointer and the program counter. The registers shown unshaded in this diagram are the ones that can be involved in load immediate instructions.

The tabulation shown in Figure 5.12(*b*) lists all those instructions available for loading eight data bits, using the immediate addressing mode, into registers B, C, D, E, H and L. In this table the first column headed 'Source code' contains the instruction in mnemonic form, the register to be loaded, a comma and the byte of data to be loaded. The second column headed 'Object code' contains the hexadecimal representation of the opcode, and also the data.

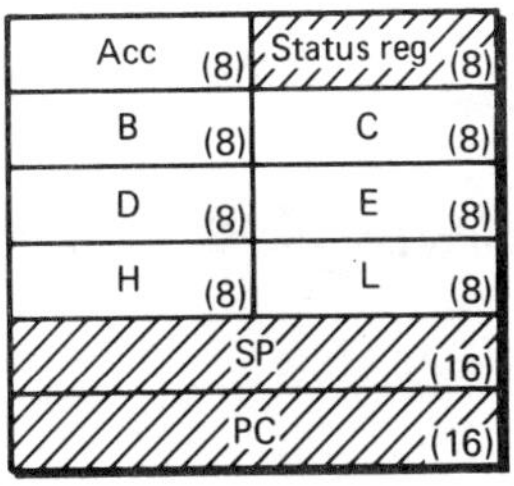

Source code	Object code
MVI A, data	3E, data
MVI B, data	06, data
MVI C, data	0E, data
MVI D, data	16, data
MVI E, data	1E, data
MVI H, data	26, data
MVI L, data	2E, data

(*a*)   (*b*)

Figure 5.12 (*a*) Register configuration of the 8085A. Unshaded registers are involved in load immediate operations (*b*) 8-bit load immediate operations

A register transfer language statement for the operation performed by the load immediate instruction is

(r) ← (byte 2)

To load register C with 4AH, the following line of program is written and stored in memory location 2000H

2000   0E 4A   MVI C, 4AH ; Move data to reg.C.

There is also a 16-bit load immediate instruction available. It consists of three bytes, where byte 1 contains the opcode, and bytes 2 and 3 hold the data which is to be loaded into a specified register pair. The unshaded registers shown in Figure 5.13(*a*) are the ones that can participate in this form of the load immediate operation.

The tabulation shown in Figure 5.13(*b*) lists all the 16-bit load immediate instructions available. The source code mnemonic for the instruction is LXI and it is followed in the tabulation by the specified register pair, a comma and the two bytes of data to be loaded. In the object code column there are three bytes for each instruction, one containing the hexadecimal representation of the opcode followed by the two data bytes.

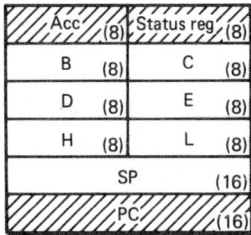

Source code	Object code
LXI B, data, data	01, data, data
LXI D, data, data	11, data, data
LXI H, data, data	21, data, data
LXI SP, data, data	31, data, data

(a)   (b)

Figure 5.13 (a) Register configuration. Unshaded registers can be used in 16-bit load immediate operations (b) 16-bit load immediate instructions

For the 16-bit load immediate instructions the register transfer language statements are

(rh) ← (byte 3)
(rl) ← (byte 2)

A line of program for loading SP with 89F0H is

2500   21 F0 89   LXI SP, 89F0H ;   Load SP

5.15 Inter-register transfers

The second group concerns inter-register transfers, where data is simply transferred directly via the internal bus system of the microprocessor from one internal register to another by a single-byte instruction. Transfer takes place from a specified source register to a specified destination register and after the execution of the instruction the contents of the source and destination registers are identical.

Most microprocessors provide instructions that allow all possible transfers between all general purpose registers. These single-byte instructions define the required operation and specify source and destination registers. Before such a transfer can be implemented the source register has to be loaded with the data to be transferred.

The unshaded registers on the register configuration diagram given in Figure 5.14(a) are the ones that can be involved in inter-register transfers, while the tabulation in Figure 5.14(b) gives the source and object codes for all possible transfers. It is worth noting that the information given in Figure 5.14(b) is normally available on the manufacturer's programming card.

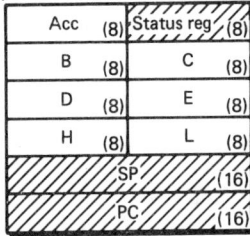

(a)

Source code	Object code	Source code	Object code	Source code	Object code
MOV A,A	7F	MOV C,C	49	MOV E,E	5B
MOV A,B	78	MOV C,D	4A	MOV E,H	5C
MOV A,C	79	MOV C,E	4B	MOV E,L	5D
MOV A,D	7A	MOV C,H	4C		
MOV A,E	7B	MOV C,L	4D	MOV H,A	67
MOV A,H	7C			MOV H,B	60
MOV A,L	7D	MOV D,A	57	MOV H,C	61
		MOV D,B	50	MOV H,D	62
MOV B,A	47	MOV D,C	51	MOV H,E	63
MOV B,B	40	MOV D,D	52	MOV H,H	64
MOV B,C	41	MOV D,E	53	MOV H,L	65
MOV B,D	42	MOV D,H	54		
MOV B,E	43	MOV D,L	55	MOV L,A	6F
MOV B,H	44			MOV L,B	68
MOV B,L	45	MOV E,A	5F	MOV L,C	69
		MOV E,B	58	MOV L,D	6A
MOV C,A	4F	MOV E,C	59	MOV L,E	6B
MOV C,B	48	MOV E,D	5A	MOV L,H	6C
				MOV L,L	6D

(b)

Figure 5.14 (a) Register configuration. Unshaded registers can take part in inter-register transfers (b) Inter-register transfer instructions

The register transfer language statement for an inter-register transfer is

$$(r_1) \leftarrow (r_2)$$

where r_2 is the source register and r_1 is the destination register.

A two-line program is given below whose function is to initially load the accumulator with data and then transfer this data to register H;

```
1000   3E A0   MVI A,A0H  ; Load accumulator
1001   67      M0V H,A    ; Transfer data from A to H
```

and the following program exchanges the contents of registers A and D, register E being used as a temporary store:

```
4000   5F   MOV E,A  ; Move contents of A to E
4001   7A   MOV A,D  ; Move contents of D to A
4002   53   MOV D,E  ; Move contents of E to D
```

There are also a number of instructions which provide the 16-bit transfers and exchanges. The registers involved in these operations are shown unshaded on the register configuration diagram in Figure 5.15(*a*) and a list of the instructions appears in Figure 5.15(*b*) with the corresponding RTL statement in the right-hand column. An examination of these statements indicates that SPHL is a transfer of data, while XCHG is an exchange of data.

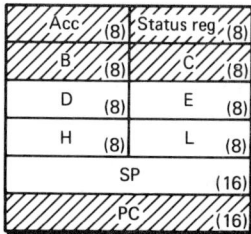

Source code	Object code	RTL statement
SPHL	F9	$(SP) \leftarrow (H)(L)$
XCHG	EB	$(H) \leftrightarrow (D)$
		$(L) \leftrightarrow (E)$

(*a*)   (*b*)

Figure 5.15 (*a*) Register configuration. Unshaded registers can be used for 16-bit transfers and exchanges (*b*) 16-bit transfer and exchange instructions.

5.16 Transfers using the direct addressing mode

The third group of instructions involved in data transfers employs the direct addressing mode. There are only a limited number of registers employed in this kind of transfer and they are shown unshaded in the register schematic diagram of Figure 5.16(*a*). A list of instructions using this method of transference is tabulated in Figure 5.16(*b*) and the RTL statements for each of these

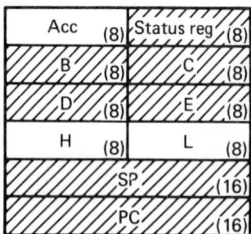

Source code	Object code	RTL statement
	8 bit transfer	
STA	32, byte, byte	$((byte\ 3)(byte\ 2)) \leftarrow (A)$
LDA	3A, byte, byte	$(A) \leftarrow ((byte\ 3)(byte\ 2))$
	16 bit transfer	
SHLD	22, byte, byte	$((byte\ 3)(byte\ 2)) \leftarrow (L)$
		$((byte\ 3)(byte\ 2)+1) \leftarrow (H)$
LHLD	2A, byte, byte	$(L) \leftarrow ((byte\ 3)(byte\ 2))$
		$(H) \leftarrow ((byte\ 3)(byte\ 2)+1)$

(*a*)   (*b*)

Figure 5.16 (*a*) Register configuration. Unshaded registers can be used for extended direct addressing transfers (*b*) Direct addressing transfer instructions

instructions are given in the last column of this table. An interpretation of the RTL statement for SHLD is, 'Store the contents of register L in the address specified by bytes 2 and 3 of the instruction and the contents of register H in the next successive memory location.' It is left to the reader to develop interpretations of the RTL statements for the other three tabulated instructions.

Since there is no direct method available for the transfer of the 16-bit contents of register pair DE, or, for that matter, BC, to memory, it can be achieved by first transferring the contents of D and E to register pair HL and from there to memory with the aid of the SHLD instruction. An assembly language program for implementing such a transfer follows:

```
2200   MOV H,D      ; Transfer contents of D to H
2201   MOV L,E      ; Transfer contents of E to L
2202   SHLD 2500H   ; Store HL in memory
```

It is possible that before implementing the above transfer that the register pair HL contains vital information which has to be saved. A modified program is shown below which saves the contents of H and L before the transfer of D and E to memory takes place. The previous program has to be modified in the following manner in order to save the contents of H and L.

```
2200   SHLD 34FEH   ; Transfer  contents  of  H  and  L  to
                      memory
2203   MOV H,D      ; Transfer contents of D to H
2204   MOV L,E      ; Transfer contents of E to L
2205   SHLD 2500H   ; Store D and E in memory
2208   LHLD 34FEH   ; Restore contents of H and L
```

An alternative and simpler program listing that achieves the same objective with two fewer instructions is given below

```
2200   XCHG         ; (DE) ⟷ (HL)
2201   SHLD 2500H   ; Store HL (was DE)
2204   XCHG         ; (DE) ⟷ (HL)
```

5.17 Transfers using the indirect addressing mode

Data transfers can also be made in the 8085A using the register indirect addressing mode. The reader may recall that a register pair is used in this mode as a pointer to the memory location containing the data or, alternatively, as a pointer to the memory location which is to receive the data. The register configuration is shown in Figure 5.17(*a*) and those registers not able to participate in this type of transfer are shown shaded. A tabulation of all the transfer instructions employing this mode is shown in Figure 5.17(*b*).

The interpretation of the RTL statement for the first seven instructions in Figure 5.17(*b*) is that 'The contents of the memory location whose address is in registers H and L is to be transferred to register r', where r can be any one of the seven unshaded registers.

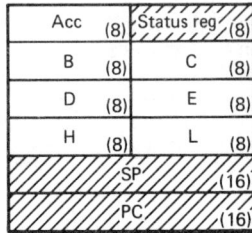

Acc (8)	Status reg (8)
B (8)	C (8)
D (8)	E (8)
H (8)	L (8)
SP (16)	
PC (16)	

(*a*)

Source code	Object code	RTL statement	Source code	Object code	RTL statement
MOV A,M	7E		MOV M,A	77	
MOV B,M	46		MOV M,B	70	
MOV C,M	4E		MOV M,C	71	
MOV D,M	56	(r) ← ((H)(L))	MOV M,D	72	((H)(L)) ← r
MOV E,M	5E		MOV M,E	73	
MOV H,M	66		MOV M,H	74	
MOV L,M	6E		MOV M,L	75	
LDAX,B	0A	(A) ← ((B)(C))	STAX B	02	((B)(C)) ← (A)
LDAX,D	1A	(A) ← ((D)(E))	STAX D	12	((D)(E)) ← (A)
			MVI M, data	36, byte	((H)(L)) ← (byte 2)

(*b*)

Figure 5.17 (*a*) Register configuration. Unshaded registers can be used for register indirect transfers (*b*) Register indirect transfer instructions

The LDAX and STAX instructions use registers, D and E as the pointer. When STAX B is implemented, the contents of the accumulator are stored in the memory location specified by the contents of the register pair BC.

The last instruction in this group is a load immediate operation and the data contained in the second byte of the instruction is transferred to the memory location specified by the contents of the register pair HL.

5.18 The input and output instructions

Finally, there are two instructions available that provide for transfers in both directions between the microprocessor and any of its associated peripherals. These two instructions are utilized in the *isolated* or *standard* I/O mode. In this mode the I/O ports are allocated an address space quite distinct from the memory address space. The instructions IN and OUT are both two-byte instructions, the first byte being the instruction opcode while the second byte contains the 8-bit address of the input or output port being accessed.

When the IN instruction appears in the program listing, the data presently held in the input port whose address is given in the second byte of the instruction is transferred to the accumulator. Alternatively, to output data, it is first taken to the accumulator and from there it can be transferred to an output port by writing the OUT instruction into the program listing. The address of the appropriate output port is given by the second byte of the OUT instruction.

A short program is given below which provides for the transfer of data 4AH from the microprocessor to the output port whose address is 05H:

```
1500   3E 4A   MVI A, 4AH   ; Move data to acc.
1502   D3 05   OUT 05H      ; Output data
```

and a second program listing is designed to input a byte of data to the accumulator and then transfer it to the memory location specified by the contents of the register pair HL:

```
1510   2A 20 15   LHLD 1500H   ; Load H and L
1513   DB 01      IN 01H       ; Input Data
1514   77         MOV M,A      ; Transfer data to memory
```

Since a port is allocated an 8-bit address a microprocessor is able to support 256 input ports and 256 output ports.

5.19 Arithmetic and logic operations

The main arithmetic and logic operations provided by the 8085A are:

(1) AND
(2) OR
(3) Exclusive-OR
(4) Addition
(5) Subtraction
(6) Compare.

Multiplication and division instructions are not normally available in the 8-bit machines; multiplication has to be implemented as a process of repeated addition while division is performed as a process of repeated subtraction. Since these methods are inherently slow an increase in the speed of these operations can only be achieved by employing peripherals specially designed to provide fast multiplication and division.

Besides the above operations, the 8085A also provides instructions for increment, decrement, complement, rotate, set and reset carry, decimal adjust accumulator, and double add.

The six basic functions listed above can be implemented using one of the three following addressing modes:

(1) Immediate
(2) Register
(3) Indirect register.

5.20 Immediate arithmetic and logic instructions

When the immediate addressing mode is utilized for executing the AND instruction, one operand is to be found in the accumulator while the second operand appears in the second byte of the immediate instruction. The AND operation, like all the main arithmetic and logical operations, is performed with eight bits. For example, if the AND instruction is to be executed, and the contents of the accumulator are 6FH while the second byte of the immediate instruction is 3AH, then the machine performs the operation shown below in the ALU and returns the result to the accumulator:

Operand 1 (Acc)	0110	1111	6F
Operand 2 (byte 2)	0011	1010	3A
	0010	1010	2A

If operand 2 is carefully chosen it is possible to selectively zero specified bits in the accumulator. To zero the four least significant bits held in the accumulator, the *masking* word or byte must be F0H. The masking operation is illustrated by the following example:

Operand 1 (Acc)	0110	1111	6F	
Operand 2 (byte 2)	1111	0000	F0	masking byte
	0110	0000	60	

In order to avoid confusion between arithmetic and logic operations, a special symbol is used to defined the AND function. Earlier in the chapter on Logic, the AND function was defined by the equation

$$f = A \cdot B$$

The dot in this equation could be misinterpreted as the symbol for multiplication, so to avoid the possibility of error it is common practice to define the AND function by the following equation.

$$f = A \wedge B$$

where the symbol $\wedge$ specifies the AND operation.

A typical example of the implementation of the OR function in an 8-bit machine using the immediate addressing mode is illustrated by the following example:

Operand 1 (Acc)	0101	1100	5C
Operand 2 (byte 2)	0010	0011	23
	0111	1111	7F

The OR operation can be used to ensure that selected bit positions in the accumulator are 'turned on'. For example, if there is a requirement that the four least significant bits in the accumulator should all be 1's then the 'turn-on' byte should be 0FH, as demonstrated in the following example:

Operand 1 (Acc)	0101	1100	5C	
Operand 2 (byte 2)	0000	1111	0F	Turn-on byte
	0101	1111	5F	

In an earlier chapter of this book, the defining equation for the OR function was given as

$$f = A + B$$

In order to avoid confusion with the arithmetic operation of addition, the OR function will from now on be defined by the equation

$$f = A \vee B$$

where the symbol $\vee$ specifies the OR function.

The Exclusive-OR operation can also be implemented using the immediate addressing mode. A typical example of this operation as performed in an 8-bit machine is given in the following example:

Operand 1 (Acc)	0101	1100	5C
Operand 2 (byte 2)	0100	1110	4E
	0001	0010	12

Selective complementation of the contents of the accumulator can be achieved with the aid of the Exclusive-OR instruction. In those bit places of the complementation byte which contain a 1, complementation of the corresponding bit in the accumulator takes place. Where the complementation byte is 0 the corresponding accumulator bit remains unchanged. An example of selective complementation follows:

Operand 1 (Acc)	0111	1110	7E
Operand 2 (byte 2)	0101	0101	55	complementation byte
	0010	1011	2B

The defining equation for the Exclusive-OR function given in the earlier chapter on Logic is

$$f = A \oplus B$$

Since the symbol $\oplus$ is unique it cannot be confused with any of the microprocessor arithmetic operations. However, there is an alternative symbol used to represent the Exclusive-OR which is given in the following equation

$$f = A \veebar B$$

For the addition of two 8-bit numbers, the first number is held by the accumulator and, assuming the immediate addressing mode, the second number will be located in the second byte of the instruction. If the addition of the two numbers produces a ninth bit, the carry flag in the status register is set to 1, otherwise it is reset to 0. Two examples of 8-bit addition follow, one in which a carry is generated while in the second example a carry is not generated.

Case 1

Operand 1 (Acc)	0101	0101	55
Operand 2 (byte 2)	1100	0001	C1
1	0001	0110	116

↑ Carry out (Carry flag set to 1)

Case 2

Operand 1 (Acc)	0101	1001	59
Operand 2 (byte 2)	0100	0001	41
0	1001	1010	9A

↑ No carry out (Carry flag reset to 0)

An alternative version of the add immediate instruction is available to deal with the situation that arises when a carry has been produced by a previous addition and has to be added in at the next stage of addition. It is possible for this to occur when multi-byte addition is being performed by the machine. If $C_{in} = 1$, the carry flag will be set to 1 and is added in during the execution of the add-with-carry instruction, as illustrated in the following example:

Carry in (C_{in})		1	01
Operand 1 (Acc)	0001	1010	1A
Operand 2 (byte 2)	0100	1010	4A
	0 0110	0101	65

↑ No carry out (Carry flag set to zero)

When a human being performs the arithmetic process of subtraction the well-established principle of borrowing from the next higher stage is employed. A microprocessor works somewhat differently and utilizes the carry flag to distinguish between positive and negative answers. For example, if the 8085A is programmed to find the difference between X = BBH (minuend) and Y = 7FH (subtrahend) the operation performed by the machine is illustrated by the following example:

Operand 1 (Acc)	X	1011	1011	BB	
Operand 2 (byte 2)	Y	1000	0001	81	(2's comp. of Y)
		1 0011	1100	3C	

↑ Carry

In this case the complement of the carry is used to reset the carry flag to zero, indicating a positive answer.

On the other hand, if the machine is programmed to find the difference between X = 5FH and Y = A3H, the subtraction is performed as shown below:

Operand 1 (Acc)	X	0101	1111	5F	
Operand 2 (byte 2)	Y	0101	1101	5D	(2's comp. of Y)
		0 1011	1100	BC	

↑ No carry

and the complement of the carry is used to set the carry flag to 1.

It will be observed that the subtrahend specified in the last calculation is greater than the minuend and the difference must therefore be negative. The absence of a carry indicates the negative nature of the answer, but unfortunately the eight bits returned to the accumulator from the ALU do not give the correct magnitude of the negative result, which is 44. The implication is that the subtraction instruction can only be usefully employed when X − Y is positive. For satisfactory manipulation of both positive and negative numbers signed arithmetic must be used.

Single-byte subtraction is best performed using the signed arithmetic mode of operation. For single-byte operation the most significant digit is used as the sign digit while the remaining seven digits represent the magnitude of the number in 2's complement form. For example, for a minuend of $(77)_{10}$ and a subtrahend of $(59)_{10}$ the subtraction is performed by the machine as illustrated in the example shown below, when the SUB instruction is executed:

$$
\begin{aligned}
(77)_{10} &= 01001101 = (4D)_{16} \\
(59)_{10} &= 00111011 = (3B)_{16} \\
\text{2's complement of } (59)_{10} &= 11000101 = (C5)_{16}
\end{aligned}
$$

Operand 1 (Acc)	0100	1101	4D	$(77)_{10}$
Operand 2 (byte 2)	1100	0101	C5	$- (59)_{10}$
(1)	0001	0010	12	$+ (18)_{10}$

↑
Carry out

Although the arithmetic operation shown above produces a carry out, when the SUB instruction is executed the carry out is complemented so that the carry status flag CY = 1.

If the value of the minuend is $(59)_{10}$ and that of the subtrahend is $(77)_{10}$ a negative answer is obtained. The sum is performed in the machine as illustrated below:

$$
\text{2's complement of } (77)_{10} = 10110011 = (B3)_{16}
$$

Operand 1 (Acc)	0011	1011	3B	$(59)_{10}$
Operand 2 (byte 2)	1011	0011	B3	$- (77)_{10}$
	1110	1110	EE	$- (18)_{10}$

and this is the 2's complement representation of $-(18)_{10}$.

There is also a subtract with borrow instruction available which utilizes the immediate addressing mode and performs the arithmetic operation $X - (Y + C_{in})$. As in the case of the add-with-carry instruction, subtract with borrow takes into account the condition of the carry flag at the beginning of the execution of the instruction. It is perhaps worth pointing out at this juncture that when writing the add with carry or subtract with borrow instructions into a program listing, the programmer should examine the value of the carry flag at the start of the instruction execution, since clearly this will affect the numerical result of either of these two operations.

For two-byte subtraction the SUB instruction is used for the least significant byte, and the subtract with borrow SBB is used for the most significant byte. An example of two-byte subtraction is shown below, and the reader will observe that it is done in two parts:

Positive no. = 005FH
Negative no. = FFA3H

For the two least significant bytes use the SUB instruction

$$\begin{array}{r} 5F \\ -A3 \end{array} \equiv \begin{array}{c} 5F \\ 5D + \\ \hline BC \end{array} \begin{array}{cc} 0101 & 1111 \\ 0101 & 1101 \\ \hline 1011 & 1100 \end{array}$$

The carry out = 0 but is complemented so that the carry flag is set to 1. For the two most significant bytes the SBB instruction is used which implements $X - (Y + Cin)$.

$$\begin{array}{r} 00 \\ -FF \\ -01 \end{array} \equiv \begin{array}{c} 00 \\ +01 \\ +FF \\ \hline (1) \end{array} \begin{array}{cc} 0000 & 0000 \\ 0000 & 0001 \\ 1111 & 1111 \\ \hline 0000 & 0000 \end{array}$$
$$\uparrow$$
Carry out

The answer is 00BCH and the carry out is complemented when the SBB instruction is executed so that the carry flag is set to 0.

Finally, the compare immediate instruction compares the magnitude of the two operands by setting or resetting the zero and carry flags according to the relative magnitudes of the two operands. When executed, the instruction subtracts its second byte from the contents of the accumulator but does not store the answer.

If X is the contents of the accumulator and Y is the second byte of the compare immediate instruction, then the operation performed is $X - Y$ and the relevant flags are set according to the following tabulation:

$X > Y$	$Z = 0$,	$C = 0$
$X = Y$	$Z = 1$,	$C = 0$
$X < Y$	$Z = 0$,	$C = 1$

The zero flag is set to 1 when $X = Y$ for the result of the subtraction is then zero, while the carry flag operates as previously described in the section on subtraction. If signed comparisons are

Source code	Object code	RTL statement
ANI, data	E6, byte	(A) ← (A) ∧ (byte 2)
ORI, data	F6, byte	(A) ← (A) ∨ (byte 2)
XRI, data	EE, byte	(A) ← (A) ∀ (byte 2)
CPI, data	FE, byte	(A) − (byte 2)
ADI, data	C6, byte	(A) ← (A) + (byte 2)
ACI, data	CE, byte	(A) ← (A) + (byte 2) + (CY)
SUI, data	D6, byte	(A) ← (A) − (byte 2)
SBI, data	DE, byte	(A) ← (A) − (byte 2) − (CY)

Figure 5.18 Immediate arithmetic and logic operations

made then it would also be necessary to examine the sign flag in addition to the zero and carry flags.

A tabulation of the six basic arithmetic and logic functions utilizing the immediate addressing mode is given in Figure 5.18, the RTL statement for each instruction being given in the right-hand column. It is not necessary to show a block schematic diagram of the registers for the immediate arithmetic and logic instructions since the only register used by these instructions is the accumulator.

5.21 Register arithmetic and logic instructions

The basic arithmetic and logic functions can also be implemented by instructions using the register addressing mode. These are all single-byte instructions and, when executed, the selected operation is performed between the contents of the accumulator and the contents of the register specified by the instruction. In addition to the six basic functions, this group contains the increment and decrement instructions which provide for the addition or subtraction of 1 from any of the general purpose registers and the accumulator. The register schematic of Figure 5.19(a) shows those registers associated with the execution of instructions in this group, while a tabulation of all the instructions in the group is shown in Figure 5.19(b), the last column in this table giving the generalized RTL statement for each operation. Since there are seven registers, A, B, C, D, E, H and L, there are seven possible instructions for each of the operations such as AND. In the object code column of the tabulation, the code for each of the seven registers is given on each row of the table, and they are listed in the order given above.

Some of the instructions in the tabulation of Figure 5.19(b) are of special interest. For example, it is possible to AND, OR or XOR the contents of the accumulator with itself. In the first two cases the contents of the accumulator remain unchanged after the

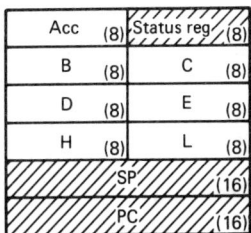

(a)

Source code	Object code for each register							RTL statement
	A	B	C	D	E	H	L	
ANA r	A7	A0	A1	A2	A3	A4	A5	$(A) \leftarrow (A) \wedge (r)$
ORA r	87	B0	B1	B2	B3	B4	B5	$(A) \leftarrow (A) \vee (r)$
XRA r	AF	A8	A9	AA	AB	AC	AD	$(A) \leftarrow (A) \forall (r)$
CMP r	BF	B8	B9	BA	BB	BC	BD	$(A) - (r)$
ADD r	87	80	81	82	83	84	85	$(A) \leftarrow (A) + (r)$
ADC r	8F	88	89	8A	8B	8C	8D	$(A) \leftarrow (A) + (r) + (CY)$
SUB r	97	90	91	92	93	94	95	$(A) \leftarrow (A) - (r)$
SBB r	9F	98	99	9A	9B	9C	9D	$(A) \leftarrow (A) - (r) - (CY)$
INR r	3C	04	0C	14	1C	24	2C	$(r) \leftarrow (r) + 1$
DCR r	3D	05	0D	15	1D	25	2D	$(r) \leftarrow (r) - 1$

(b)

Figure 5.19 (a) Register configuration for the 8085A. Unshaded registers can be used for 8-bit register-based arithmetic and logic operations (b) 8-bit register arithmetic and logic instructions

operation but it sets the flags as relevant, in particular the carry flag is always set to zero. If the XOR operation is performed, the accumulator is cleared, as the following example illustrates:

Operand 1 (Acc)	1011	0111	B7
Operand 2 (Acc)	1011	0111	B7
	0000	0000	00

The reader will also observe that the accumulator is always a part of the instruction, so it is not possible to do a direct AND, OR or XOR of the D and H registers, or any other pair of registers that does not include the accumulator. To implement the function D ∨ H the contents of D must first be transferred to the accumulator and then the contents of A can be ORed with the contents of H, as the following program listing illustrates:

MOV A, D ; Transfer contents of D to A
ORA H   ; OR contents of H and A

After the execution of ORA, registers D and H contain the original data while the accumulator contains the result of the required operation.

In the arithmetic group of instructions perhaps the one which is of most interest is the addition of the contents of the accumulator to itself. When this instruction is executed the contents of the accumulator are shifted one place to the left, which is equivalent to multiplying them by two. The example given below illustrates this point:

Operand 1 (Acc)   0100 1011   4B
Operand 2 (Acc)   0100 1011   4B
                   ---- ----   ----
                   1001 0110   96 = 4B × 2
                   ← Shift left

Facilities are also not provided for adding, subtracting, or comparing directly the contents of a pair of registers such as E and L. To perform any one of these operations the contents of E have first to be transferred to the accumulator. If, for example, a comparison is now required, then it can be made between the contents of the accumulator and register L. A short assembly language program to perform this operation follows:

MOV A, E   ; Transfer contents of E to A
CMP L   ; Compare A with L

The instructions dealt with previously in this section implement 8-bit arithmetic and logic operations. There are also three instructions employing the register addressing mode associated

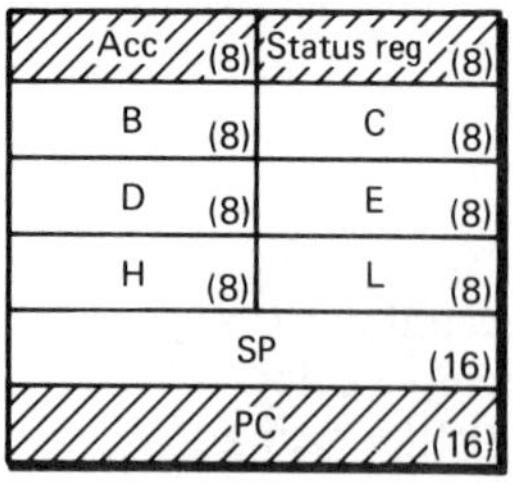

Source code	Object code	RTL statement
DAD B	09	
DAD D	19	(H) (L) ← (H) (L) + (rh) (rl)
DAD H	29	
DAD SP	33	
INX B	03	
INX D	13	(rh) (rl) ← (rh) (rl) + 1
INX H	23	
INX SP	33	
DCX B	0B	
DCX D	1B	(rh) (rl) ← (rh) (rl) − 1
DCX H	2B	
DCX SP	3B	

(a)

(b)

Figure 5.20 (a) Register configuration for the 8085A. Unshaded registers can be used for 16-bit register-based arithmetic and logic operations (b) 16-bit register arithmetic and logic instructions with register pairs that perform the increment, decrement and add operations. Because these instructions are associated with register pairs they provide 16-bit operations. The registers involved in these 16-bit operations are identified in the register schematic diagram (Figure 5.20(a)) and a tabulation listing the operations available appears in Figure 5.20(b).

An examination of the RTL statement for the DAD instruction indicates that it gives a direct implementation of 16-bit or double-length addition. The H and L registers may be regarded as a 16-bit accumulator to which the contents of register pairs BC, DE, HL, or SP may be added. However, after the execution of DAD the status flags are unchanged with the exception of the carry flag.

5.22 Indirect register arithmetic and logic instructions

Finally, there is a group of arithmetic and logic instructions which employ indirect register addressing. These instructions have the first operand located in the accumulator while the second operand is to be found at a memory address specified by the contents of the H and L registers. For example, the instruction XRA M can be interpreted as an Exclusive-OR operation between the contents of the accumulator and the contents of the register at the memory address specified by the contents of the H and L registers.

It is unnecessary to show a register schematic for this group of instructions because the only register that can participate in the arithmetic and logical operations using this addressing technique is the .accumulator. However, a list of arithmetic and logic instructions available employing this technique is listed in Figure 5.21.

Source code	Object code	RTL statement
ANA M	A6	$(A) \leftarrow (A) \wedge ((H)(L))$
ORA M	B6	$(A) \leftarrow (A) \vee ((H)(L))$
XRA M	AE	$(A) \leftarrow (A) \veebar ((H)(L))$
CMP M	BE	$(A) \leftarrow ((H)(L))$
ADD M	86	$(A) \leftarrow (A) + ((H)(L))$
ADC M	8E	$(A) \leftarrow (A) + ((H)(L)) + (CY)$
SUB M	96	$(A) \leftarrow (A) - ((H)(L))$
SBB M	9E	$(A) \leftarrow (A) - ((H)(L)) - (CY)$
INR M	34	$((H)(L)) \leftarrow ((H)(L)) + 1$
DCR M	35	$((H)(L)) \leftarrow ((H)(L)) - 1$

Figure 5.21 Indirect register arithmetic and logic instructions

As an example of the use of the indirect register arithmetic and logic instructions, a program listing is given below which adds the contents of memory location 5000H to the contents of the accumulator and stores the result in register C. This operation is followed by the ANDing of the contents of the accumulator with the contents of 5000H and the result is stored in register D. Finally the contents of 5000H are cleared.

2000	21 00 50	LXI H, 5000H	; Load address in H & L registers
2003	3A 01 50	LDA 5001H	; Load acc.
2004	86	ADD M	; Add contents of A to contents of memory location specified by HL
2007	4F	MOV C,A	; Move A to C
2008	A6	ANA M	; AND contents of A to contents of memory location specified by HL
2009	57	MOV D,A	; Move A to D
200A	AF	XRA A	; Clear Accumulator
200B	77	MOV M,A	; Move contents of A to memory location specified by HL.
200C	76	HLT	; Stop

5.23 Rotate and data shift instructions

Data rotation and shift is an operation that finds many applications in a program listing. For example, the process of multiplication can be described as a shift and add process, while division by the same token is a shift and subtract process. The shift instruction will be a feature of the program listing for either of these two operations.

There are a number of forms of the shift operation, the simplest one being the straightforward shift without rotation, as illustrated in Figure 5.22. The register is holding the data ABH and the most

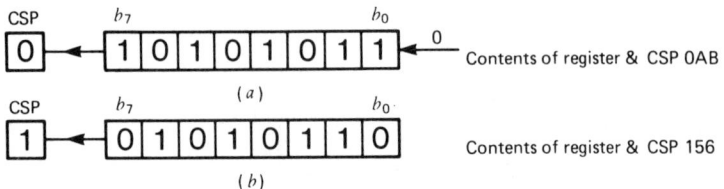

Figure 5.22 The left shift (a) Before shift (b) After shift

significant bit of the register is connected to the carry status position (CSP) in the status register. If a 0 is injected into the least significant position of the register, all bits of higher significance move one place to the left and the most significant bit is transferred to the carry status position. After the execution of the shift, the contents of the register become 56H and the carry status bit becomes 1, as indicated in Figure 5.22(b). The shift left has, in effect, resulted in the multiplication by 2 of the original contents of the register since AB $\times$ 2 = 156H, the number now appearing in the CSP and register. Two left shifts would result in multiplication by 4 while n left shifts would result in multiplication by 2^n.

For a right shift, a 0 is injected into the most significant end of the register and the least significant bit is connected to the carry status position. If, initially, the register contains 42H, as shown in Figure 5.23, then, after the shift has been executed, the register considered in conjunction with the carry will contain 21H. The right shift has resulted in division by 2. It follows that two right shifts will result in division by 4, and so on.

If a circular rotation of the contents of the register is required in either direction, then the most significant position of the register is connected to its least significant position. For a left circular rotation b_7 is also connected to the carry status position so that when a left circular shift is executed, the most significant bit b_7 is transferred to bit position b_0 and also to the carry status position, while all other bits in the register move one place left. For a right circular rotation, the least significant bit b_0 is transferred to bit

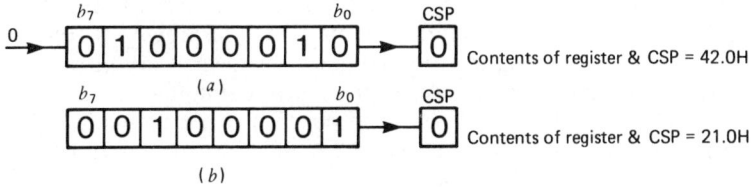

Figure 5.23 The right shift (a) Before shift (b) After shift

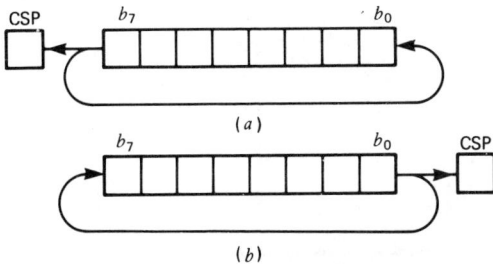

Figure 5.24 Circular rotation (a) Left circular rotation (b) Right circular rotation

position b_7 and also to the carry status position, whilst all other bits in the register move one place right. These two operations are illustrated in Figure 5.24. In either case, after eight consecutive shifts have been executed the data in the register will have returned to its original position.

An alternative process to the one described above is circular rotation through carry, which is described in Figure 5.25 for both left and right rotation. With this operation, nine consecutive shifts

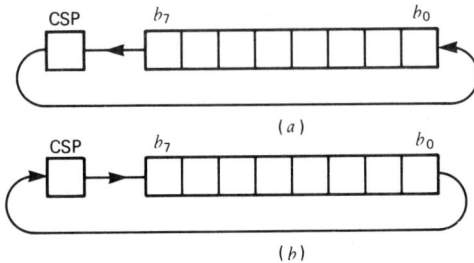

Figure 5.25 Circular rotation through carry (a) Right rotation (b) Left rotation

would have to take place before the register contents and the carry status bit returned to their original condition.

In the 8085A, two of these operations are provided. They are: (a) rotate left or right; (b) rotate through carry left or right. This results in the provision of four separate instructions which are tabulated in Figure 5.26, where the right-hand column gives the RTL statement for each instruction. In the case of the RLC instruction, the RTL statement may be interpreted as follows. Generalizing, bit A_n in the accumulator will take up bit position A_{n+1}; in particular, bit A_7 is transferred simultaneously to bit position A_0 and to the carry status position. An examination of all the rotate and shift instructions in Figure 5.26 shows that they are provided for the accumulator only.

Source code	Object code	RTL statement
RLC	07	$(A_{n+1}) \leftarrow (A_n); (A_0) \leftarrow (A_7)$ $(CY) \leftarrow (A_7)$
RRC	0F	$(A_n) \leftarrow (A_{n+1}); (A_7) \leftarrow (A_0)$ $(CY) \leftarrow (A_0)$
RAL	17	$(A_{n+1}) \leftarrow (A_n); (CY) \leftarrow (A_7)$ $(A_0) \leftarrow (CY)$
RAR	1F	$(A_n) \leftarrow (A_{n+1}); (CY) \leftarrow (A_0)$ $(A_7) \leftarrow (CY)$

Figure 5.26 The rotate and shift instructions for the 8085A

An application of the use of the rotate accumulator right instruction RRC is given in the word disassembly program listed below. The data stored in memory location 3000H is 5AH and the purpose of the program is to disassemble the stored data so that 0AH is stored in location 3001H and 05H is stored in location 3002H.

2000	21 00 30	LXI H, 3000H	; Load address in H and L registers
2003	7E	MOV A,M	; Move contents of address specified by HL to A
2004	57	MOV D,A	; Move contents of A to D
2005	E6 0F	ANI 0FH	; Mask off 4 most significant digits
2007	23	INX H	; Increment H
2008	77	MOV M,A	; Transfer A to memory address specified by HL
2009	7A	MOV A,D	; Transfer D to A
200A	0F	RRC	; ⎤
200B	0F	RRC	; ⎟ Shift A right
200C	0F	RRC	; ⎟ four places
200D	0F	RRC	; ⎦
200E	E6 0F	ANI 0FH	; Mask off 4 most significant digits
2010	23	INX H	; Increment H
2011	77	MOV M,A	; Transfer A to memory location specified by HL
2012	76	HLT	; Stop

5.24 Miscellaneous instructions

The carry bit in the status register is always set to 0 after the execution of any instruction involving the three logic operations AND, OR and Exclusive-OR. However, after the implementation of any arithmetic instruction the carry bit is set to 0 or 1 depending upon the result of the operation.

Additionally, there are two instructions the programmer can invoke if a change in the value of the carry status is required. The CMC instruction complements the carry status, and hence can be used to reset the carry to zero if required, while the STC instruction, when executed, always sets the carry flag to 1 irrespective of its previous value.

When a microprocessor adds or subtracts, it does so in the binary number system. For example, two 8-bit words representing the BCD numbers 09 and 07, when added in the arithmetic/logic unit of the machine, give a BCD answer of 10 rather than the required 16, as illustrated below:

```
0000   1001   09
0000   0111   07
0001   0000   10   Incorrect answer
```

It has been shown in the chapter on arithmetic that to correct the above answer 06 has to be added in, and the introduction of this correction factor is shown below:

```
0001   0000
0000   0110
0001   0110   16
```

Hence, when implementing BCD addition in the machine, the ADD instruction has to be followed by the decimal adjust instruction DAA, which simply modifies the incorrect answer into the correct BCD form.

If a BCD number appears in the accumulator and is to be added to a BCD number stored in a memory location specified by the contents of the H and L registers, the following instruction sequence is required:

```
ADD M    86
DAA      27
```

The rules for applying the correction when a BCD addition is being performed are:

(1) If the value of the least significant four bits in the accumulator is >9, or if there is a carry from bit 3 to bit 4, 6 is added to the contents of accumulator.
(2) If the value of the most significant four bits in the accumulator is >9, or if the carry flag is set, 6 is added to the four most significant bits.

It is clear that to implement the first of these rules the machine has to be aware of the presence of a carry from bit 3 to bit 4, so that the correction can be applied when the DAA instruction is executed. For this reason the carry referred to as the auxiliary carry AC is stored in the status register after the execution of the ADD instruction.

The DAA instruction cannot be used after a BCD subtract operation. The correct way to do BCD subtraction with the 8085A is to take the 9's complement and then add. For example, to subtract the contents of register C from register B the following program listing would be used:

```
MVI   A, 99H
SUB   C          ; Form 9's complement
ADD   B          ; Add
DAA              ; Decimal adjust
```

Problems

5.1   State the kind of addressing used in each of the following 8085A instructions, and justify your answer

(a) ANA   M   (d) XCHG
(b) CMP   r   (e) PUSH   rp
(c) XRI   data (f) OUT   port

Give the hexadecimal equivalents and also a register transfer language statement for each of these instructions.

Draw a block diagram of the registers used when executing ANA   M and show how data is moved about in the machine during the implementation of this instruction.

5.2   (a) Find the effective address of an instruction, if the contents of the index register are 0397H while the address offset is 78H.
(b) What should be the value of the address offset if the effective address is (i) 0413H and (ii) 0649H ?
(c) If the address offset is represented in the signed 2's complement convention, determine the effective address if the contents of the index register are 2CF0H and the address offset is (i) 7AH and (ii) 9CH.

5.3   Using two different programming techniques, write program listings for obtaining the 2's complement of any number stored in memory location 0100 and return the result to memory location 0150H.

5.4   If register A contains 5CH, register C contains 49H and the contents of registers H and L are 1000H, implement each of the following instructions in turn. After implementation, determine the contents of registers A, C, H and L and the condition of the flags:

(1) MOV   M,C   (7) ANA   A
(2) MVI   A, F2H (8) XRI   29H
(3) INR   A   (9) CMP   C

(4) SUI   39H   (10) CPI   34H
(5) SUB   M   (11) RRC
(6) ADD   C   (12) CMA

5.5   Draw a block diagram incorporating all the registers and the ALU of the 8085A and both data and program memory. Show the execution of the following instructions on your block diagram

(a) CPI   B7H
(b) LDAX   B
(c) SBI   29H
(d) LHLD

If the accumulator contains 49H before the execution of CPI and SBI, determine its contents and the condition of the flags after execution.

5.6   Give an example of (a) a one-byte instruction, (b) a two-byte instruction and (c) a three-byte instruction available with the 8085A, and describe the use of the instructions you have selected with a short program listing.

5.7   If the contents of the accumulator are 72H, what is the effect of executing the following instructions on the contents of the accumulator and the flags?

(a) XRA   A
(b) ANI   0FH
(c) ORI   80H
(d) ORA   A
(e) ADD   A

6 The instruction set II: jump, call, return and stack control

6.1 Introduction

The simple program sequences introduced in the last chapter have consisted of a list of instructions that are processed consecutively. For example, if the program counter is initially set to 0000H and the terminating instruction HLT is located at 00A5H, it sequences through the 166 consecutive memory locations and when memory location 00A5H is reached program execution terminates. All instructions in the program sequence are executed, none is omitted, none is repeated. However, in certain circumstances some sections of a program sequence have to be omitted, and in some cases a certain section of the program may have to be repeated a number of times. This facility is provided by the inclusion of JUMP instructions in the instruction set and these may be unconditional or conditional. The use of a JUMP instruction implies that some method has to be devised for changing the contents of the program counter at the point in the program sequence where the JUMP instruction is encountered.

It is also useful to devise subroutines, which are stored in memory and which may be used more than once within the main program sequence. Commonly used sequences can be developed as subroutines and occupy much less memory space than would otherwise be employed if the sequence is written into the main program every time it is required. It is now common practice to establish a library of subroutines that implement common functions. Clearly, subroutines can then be drawn from the library and inserted into a wide variety of programs.

To enter a subroutine, the machine has to divert from the main program sequence and at the end of the subroutine it has to return to the original sequence at the point of diversion. These facilities are provided by the CALL and RETURN instructions. The CALL instruction is part of the main program sequence while the RETURN instruction is the last in the subroutine sequence. In either case these instructions may be conditional or unconditional.

When the machine diverts from the main program sequence and enters a subroutine, its starting address has to be inserted into the program counter whose original contents have to be stored in a temporary storage location. The portion of memory allocated to the stack is used for the temporary storage of this type of data. At the end of the subroutine the original contents of the program counter are returned from the stack and reinserted into the program counter. Transfer of data to and from the stack is achieved by the PUSH and POP instructions, which effectively control stack operations.

In this chapter the control instructions JUMP, CALL, RETURN, and those governing stack operations, will be examined in some detail. Attention will be paid to the conditions that can be imposed on the execution of the JUMP, CALL and RETURN instructions. The application of conditions to these instructions implies that the machine has a capacity to make simple decisions; this ability can be illustrated with the aid of flowcharts. Consequently, a description of the basic elements of flowcharts and their uses will be given first.

6.2 Flowcharts

Flowcharts are used to give a pictorial illustration of the flow and structure of a program. There are three main building blocks used in flowcharts. They are:

(a) The terminal blocks, as illustrated in Figure 6.1(a), which are used to identify the beginning and end of a program or subroutine. The starting block contains the name of the

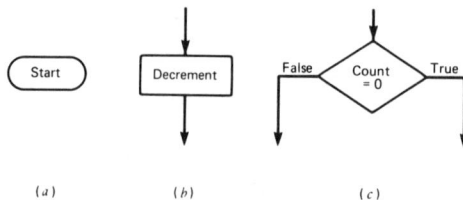

(a)   (b)   (c)

Figure 6.1 Flowchart building blocks (a) Terminal block (b) Process block (c) Decision block

program or subroutine or, alternatively, the word START, while the terminating block will contain the word END or, in the case of a subroutine, RETURN.

(b) A process block (Figure 6.1(*b*)), which represents a definable task that has to be performed within the main program sequence. It is then normal procedure to indicate the nature of the activity by inserting a phrase or a word within the block, such as DECREMENT or INITIALIZE. The activity may be complex and require a number of instructions for its implementation, or it may be an elementary process which can be implemented by a single instruction.

(c) A decision block (Figure 6.1(*c*)), represented by a diamond. It has one input and two possible outputs. The selection of the output path is governed by the condition inserted in the block. This condition can be represented by a logical proposition which may be true or false, and each of these two possibilities is associated with one of the output paths.

A typical example of a flowchart is shown in Figure 6.2. The flowchart chosen is one that describes the shift and add multiplication process. The first block A, labelled 'initialize',

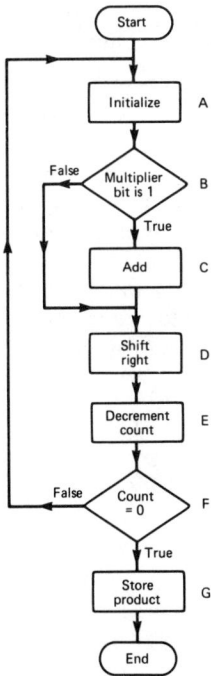

Figure 6.2 Flowchart for a multiplication routine

specifies the initial conditions which must be set up before multiplication can be commenced. In this block the contents of the accumulator are set to 00H, the multiplicand and multiplier are stored in specified locations in data memory and a count is set up in a general purpose register that equals the number of right shifts to be performed.

Block B in Figure 6.2 is a decision box from which there are two exits. If the multiplier bit is 1 the proposition in the block is true and the next operation appears in block C where the add process is specified, the multiplicand being added to the contents of the accumulator to form the next partial product. Alternatively, if the multiplier bit is 0 the proposition is false and a jump is required which omits block C. The two paths rejoin one another and enter the process box D where the shift operation takes place.

Having now performed either a shift and add or, alternatively, a shift only operation, the number stored in the count register has to be decremented. This operation is carried out in block E. After decrementing the count, decision block F is entered. If the count is not zero the proposition in the block is false, a jump is executed and the loop containing blocks B C D E F is traversed again. Alternatively, if the count is zero, the proposition is true, block G is entered and the product is stored in the appropriate memory location. This process completed, the program is terminated.

6.3 Basic flowchart combinations

The process and decision boxes can be connected in a number of different ways, but in practice, in order to produce programs that are well structured and easily understood, only a limited number of combinations of the basic flowchart blocks are used.

The simplest combination of the basic building blocks is the SEQUENCE structure shown in Figure 6.3. This represents a subroutine or process followed sequentially in time by a second

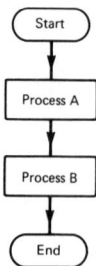

Figure 6.3 The SEQUENCE structure

subroutine or process. It is, in essence, a program sequence in which no decisions have to be made by the machine.

A second commonly used combination of the basic building blocks is the IF-THEN/ELSE structure, illustrated in Figure 6.4(*a*). This structure allows the selection of one of two processes, depending on the truth or falsity of the proposition appearing in the decision box. If the proposition is false, process A will be implemented. If it is true, process B will be selected. The IF-THEN/ELSE structure can be defined by the statement 'IF the specified proposition is true then implement process A or ELSE implement process B.'

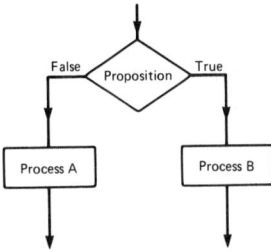

(a)

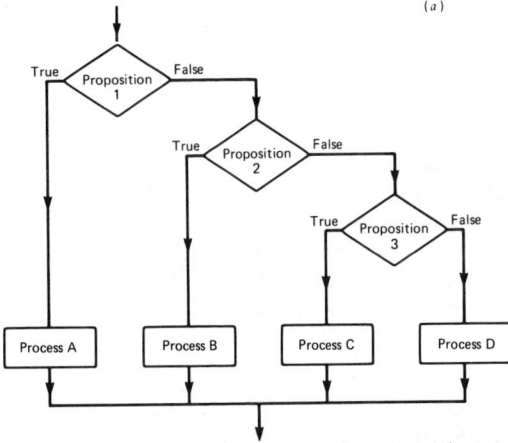

(b)

Figure 6.4 (*a*) The IF-THEN/ELSE structure (*b*) A combination of three IF-THEN/ELSE structures forming a SELECT structure

It is possible to generate a SELECT structure by combining a number of IF-THEN/ELSE structures. A combination of three of them is shown in Figure 6.4(*b*). The resulting structure allows the selection of one out of four processes – A, B, C or D.

There are also two basic structures available for providing the repetition of a process a number of times:

(1) The DO-WHILE structure (Figure 6.5(*a*)), where the condition specified by the proposition in the decision box must be tested after each repetition of process R. As long as the proposition is true, further repetitions of process R occur. When the condition changes and the proposition becomes false, the alternative exit is taken from the decision box and no more repetitions are initiated. This structure may be defined by the statement 'DO process R WHILE specified proposition is true.'

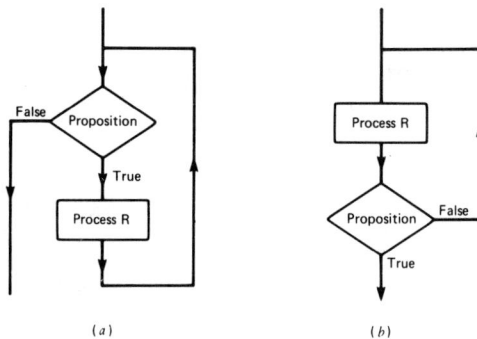

(*a*)   (*b*)

Figure 6.5 (*a*) The DO-WHILE structure (*b*) The DO-UNTIL structure

(2) The DO-UNTIL structure described in Figure 6.5(*b*) is a repetitive process structure where the condition which is tested by the proposition changes after each repetition. When the proposition is true the alternative exit is taken from the decision box and the repetitions cease. The structure is defined by the statement 'DO process R UNTIL the proposition is true.'

The structures described in this section constitute the basic building blocks for flowcharts and suitable combinations of them can be used to define any program. One of the main advantages of these structures is that they have single entry and exit points, and it follows that combinations of them will also possess the same advantage. This leads to the technique of constructing a program from a number of structural elements which can be developed and tested independently and, if necessary, replaced without altering the remaining units. Programs developed in this way are more readily understood and usually more reliable.

Flowcharts constructed in the way described above are called *structured flowcharts* and the corresponding program is called a *structured program*.

6.4 The JUMP instructions

It is characteristic of all JUMP instructions, irrespective of whether they are conditional or unconditional, that their execution will result in a change of contents of the program counter. All of them, with the exception of PCHL, are three-byte instructions, the first byte being the instruction opcode, and bytes 2 and 3 providing the address to which the jump has to take place, thus identifying where the next instruction will be found.

The jump can be either forward or backward, as illustrated in Figure 6.6. The instruction opcode C3H specifies an unconditional jump in both cases. In Figure 6.6(*a*) the jump is forward, the forward address 2000H being supplied by bytes 2 and 3 of the jump instruction, while in Figure 6.6(*b*) a backward jump is illustrated, the backward address 1000H being given by bytes 2 and 3 of the instruction. The new address in both cases is transferred to the program counter whose original contents are lost.

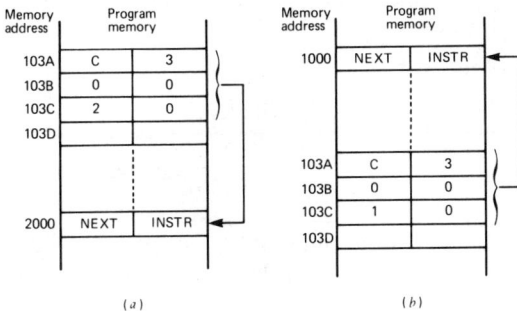

Figure 6.6 (*a*) The forward jump (*b*) The backward jump

A simple example of the use of the instruction JMP is the generation of an endless loop which could be used to terminate the execution of a program. The last line of the program would then read as follows

159A   C3 9A 15   LOOP   JMP   LOOP;

The reader should notice the use of the label LOOP. The mnemonic for jump, JMP, is written, followed by the label LOOP, which then indicates that the next instruction after the execution of JMP will be preceded by the label LOOP and hence is easily identified in the program listing. In the above case, after the execution of JMP the program returns to JMP and continues to do so after every execution of that instruction. The machine is locked in an endless loop.

Conditional jumps are initiated after an examination of one of the four status flags, i.e. carry, zero, parity and sign. If the specified condition is true, control is transferred to the instruction whose address is specified in bytes 2 and 3 of the conditional jump instruction. A tabulation of all the jump instructions, both conditional and unconditional, is given in Figure 6.7. Two additional columns have been added to the instruction table giving the required condition and its corresponding flag condition for the implementation of the jump.

Source code	Object code	Condition	Flag condition	RTL statement
JMP, addr	C3	None	None	(PC) ← (byte 3)(byte 2)
JC, addr	DA	Carry	C = 1	
JNC, addr	D2	No carry	C = 0	
JM, addr	FA	Minus	S = 1	
JP, addr	F2	Plus	S = 0	(PC) ← (byte 3)(byte 2)
JPE, addr	EA	Even parity	P = 1	if specified
JPO, addr	E2	Odd parity	P = 0	condition is met
JZ, addr	CA	Zero	Z = 1	
JNZ, addr	C2	Not zero	Z = 0	
PCHL	E9	None	None	(PCH) ← (H) (PCL) ← (L)

Figure 6.7 Jump instructions for the 8085A

It is clear that the execution time of the conditional jump instruction will have two distinct values. If the specified condition is not met, the program sequence continues without interruption. However, if the condition is met then interruption of the program sequence occurs and bytes 2 and 3 of the instruction are transferred to the program counter. In the first case seven clock periods are required for execution, while if the jump is implemented ten clock periods are needed.

The last instruction PCHL in Figure 6.7 is a single-byte instruction which transfers the contents of the H and L registers to the program counter. It may be regarded as a register indirect jump instruction in that the H and L registers contain the address to which the jump has to be made.

6.5 The program implementation of the IF-THEN/ELSE structure

Conditional jump instructions in a program sequence indicate the presence of a decision box in the corresponding flowchart. Since most of the basic flowchart combinations discussed earlier in this chapter include decision boxes it follows that their conversion to a program sequence will require the inclusion of conditional jump instructions in that sequence. For example, suppose the task to be performed is to compare two numbers stored in specified memory locations, determine which of them is the larger, and then store it in some other specified memory location. A flowchart describing this problem is shown in Figure 6.8, and it will be noticed that it contains the IF-THEN/ELSE structure.

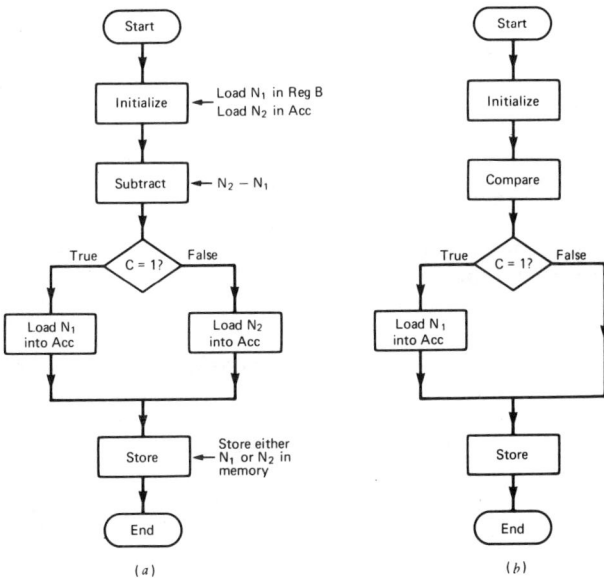

Figure 6.8 (a) Flowchart for determining the larger of two numbers using the subtract instruction (b) As for (a) except that the COMPARE instruction is used rather than subtraction

The initialize block allows for N_1 to be stored in a general purpose register and N_2 in the accumulator. After initialization, N_1 is subtracted from N_2, and if after the completion of this operation the carry $C = 0$, then $N_2 \geqslant N_1$, while if $C = 1$, $N_2 < N_1$. The condition of the carry flag leads to the selection of one of two processes and the appearance of the IF-THEN/ELSE structure in the flow chart. The interpretation of the structure is that IF $C = 1$ THEN load N_1 into the accumulator. ELSE load N_2 into the accumulator. At the termination of the selected process the contents of the accumulator are stored in memory.

If N_1 is stored in memory location 2040H and N_2 is stored in memory location 2041H, the following assembly language program when executed will determine the larger of the two numbers and store it in memory location 2042H:

```
            LDA    2040H   ;  Load N₁ in Acc
            MOV    B,A     ;  Transfer N₁ to B
            LDA    2041H   ;  Load N₂ in Acc.
            SUB    B       ;  N₂−N₁
            JC     MORE    ;  If C = 1 jump
            ADD    B       ;  Add B to Acc.
            JMP    NEXT    ;
MORE        MOV    A,B     ;  Transfer N₁ to Acc.
NEXT        STA    2042H   ;  Store data in memory
            HLT            ;
```

The above problem could have been solved by replacing the SUB instruction with the compare (CMP) instruction. After the compare instruction has been executed the contents of the accumulator remain unchanged and it will still hold the number N_2. This leads to a modification of the flow chart (Figure 6.8(b)) in that the process box in the right-hand branch of the IF-THEN/ELSE structure is no longer needed and the structure itself is modified to an IF-THEN structure. The modified program using the compare method is given below:

```
            LDA    2040H   ;  Load N₁ in Acc.
            MOV    B,A     ;  Transfer N₁ to B
            LDA    2041H   ;  Load N₂ in Acc.
            CMP    B       ;  Compare N₂ with N₁
            JC     MORE    ;
            JMP    LESS    ;
MORE        MOV    A,B     ;  Transfer N₁ to Acc.
LESS        STA    2042H   ;  Store data in memory
            HLT            ;
```

6.6 Program loops

When a sequence of program instructions is traversed repeatedly, a program loop has been formed. At the end of the sequence a conditional jump instruction has to be written so that control can be returned to the beginning of the sequence.

One simple looping technique is to set up a count in one of the microprocessor general purpose registers; after each traversal the count is decremented and examined. When the value of the count is zero an exit is made from the loop and the main program sequence is rejoined. In this case the number of times the loop is to be repeated is known before the loop is entered. An alternative method is to count the number of repetitions of the loop and a test can be made for the required count at the end of each traversal.

Repetitive loops of this nature can be described on a flowchart by the DO-WHILE and DO-UNTIL structures. As an example of the formation of a repetitive loop, programs will be developed using these two structures to determine the sum of four numbers stored in consecutive memory locations, the first of the four numbers being stored at 1000H. After the summation is complete, the total is to be stored in memory location 1010H.

A flowchart for the DO-UNTIL description of this problem is shown in Figure 6.9(a). In the initialize block the accumulator is cleared, a count of 4 is set up in, say, register B and a pointer to the memory location 1000H is set up in the H and L registers. The contents of memory location 1000H are added to the contents of the accumulator and the count is decremented. The add and decrement loop is repeated until the count is zero. When that condition is met the main program sequence is rejoined and the sum is stored in memory. The program corresponding to this flowchart structure follows:

```
        LXI    1000H   ;  Load H and L registers
        MVI    B,04H   ;  Set up count in B reg.
        XRA    A       ;  Clear Acc.
LOOP    ADD    M       ;  Add
        INX    H       ;  Increment H and L registers
        DCR    B       ;  Decrement count
        JNZ    LOOP    ;
        STA    1010H   ;  Store Accumulator
        HLT
```

The same problem is described in Figure 6.9(b) by the DO-WHILE structure. The basic difference between the two structures is that in Figure 6.8(a) addition takes place before the count is decremented, while in Figure 6.8(b) the count is decremented first, followed by the addition. In the first case the process can be described as execute and count, while in the second

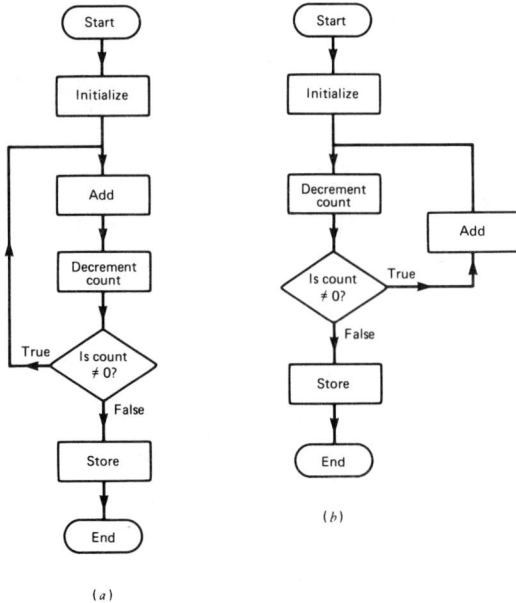

Figure 6.9 Repeated addition (a) Using the DO-UNTIL structure and (b) using the DO-WHILE structure

case it is count and execute. There is however, one subtle difference between the two cases. For the DO-UNTIL structure the initial count set up in register B is 4. However, in the DO-WHILE structure the count is decremented before the first addition; consequently, to add four numbers requires an intial count of five in register B. The assembly language program for the DO-WHILE description of the problem is given below.

```
        LXI   1000H  ; Load H and L registers
        MVI   B,05H  ; Set up count in B reg.
        XRA   A      ; Clear Acc.
LOOP    DCR   B      ; Decrement count
        JZ    DONE   ;
        ADD   M      ; Add
        INX   H      ; Increment H and L registers
        JMP   LOOP   ;
DONE    STA   1010H  ; Store Acc.
        HLT
```

6.7 A program implementation of the SELECT structure

The SELECT structure was earlier shown as a combination of IF-THEN/ELSE structures. However, there is an alternative method of selecting one out of a number of processes, using the PCHL instruction.

For example, suppose that one out of six processes has to be selected, each one of these processes being represented by a sequence of instructions stored in memory. It is convenient to store the starting address of each process in a table, as illustrated in Figure 6.10(a) where the starting address of the table is 2000H. It will be observed that the starting address for each of the required processes is given by bytes 2 and 3 of a series of

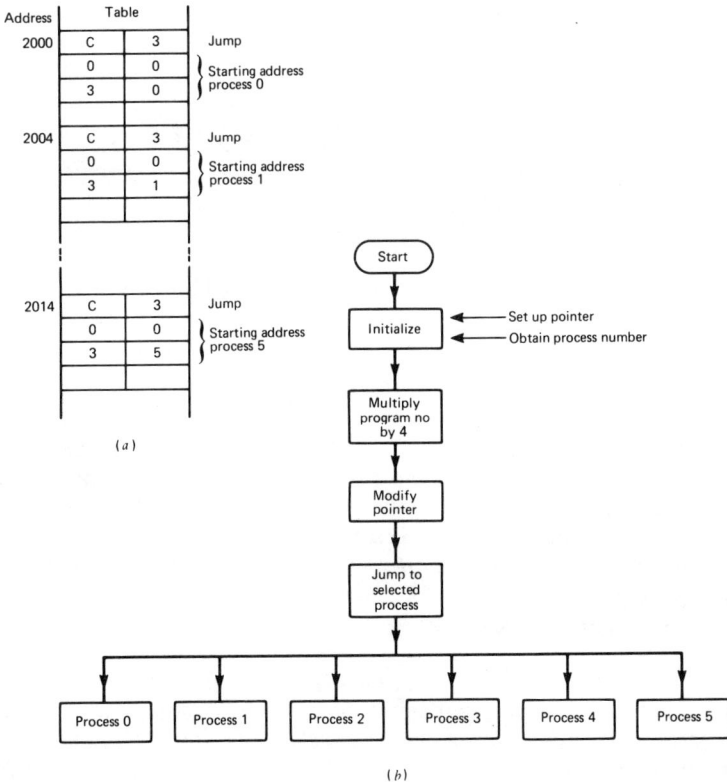

Figure 6.10 (a) Table providing the starting addresses for six processes (b) Flowchart for selection of one out of six processes

unconditional jump instructions, the starting address of process 1 being given by 3100H. In the main program sequence it is now necessary to develop a method for establishing a branch to the starting address of any one of the six processes identified in the table.

The selection of one of the six processes is described by the flowchart (Figure 6.10(b)). In the initialize block the starting address of the table is set up in the H and L registers, while a number identifying the selected process is loaded into the accumulator. For process 1, the appropriate number is 01H. If this number is now multiplied by 4 then the contents of the accumulator become 04H. This number is now written into the L register thus modifying the address previously established in the H and L registers, the new value of the address being 2004H which is the required starting address for process 1. Transference of this address to the program counter is now achieved by writing the PCHL instruction which places the contents of the H and L registers in the program counter.

The program corresponding to the flow chart is given below:

```
LXI    H 2000H    ;  Load starting address of table
MVI    A 01H      ;  Obtain process no.
ADD    A          ;  Multiply by 2
ADD    A          ;  Multiply by 2
MOV    L,A        ;  Modify address to L reg.
PCHL              ;  Insert address into PC
```

6.8 Stack operations and instructions

When a subroutine is called, its starting address is transferred to the program counter. The machine sequences through the list of instructions which constitute the subroutine, and at the end of the routine the machine has to return to the next instruction in the main program sequence. This means that in effect the contents of the program counter at the point of entry to the subroutine have to be saved and then returned to the program counter at the end of the routine.

The stack is used for storing this kind of information on a temporary basis. It can consist of a set of registers within the microprocessor chip, or, alternatively, a section of read/write memory can be set on one side by the programmer for stack operations. In both cases the stack has a last-in, first-out (LIFO) structure, which implies that the last word placed on the top of the stack will be the first one off the stack when the reverse process takes place.

A stack having a depth of eight registers is shown in Figure 6.11. When the first word, W_1, is entered onto the stack it occupies the top register. On the arrival of W_2 it displaces W_1, which moves to the next lower register in the stack, and when W_3 arrives it occupies the top register and displaces W_2 and W_1 into their next lower registers respectively. In this example, a maximum of eight words can be stored on the stack and it is essential that the programmer ensures that this limit is not execeeded, otherwise information is lost from the bottom register, resulting in machine malfunction.

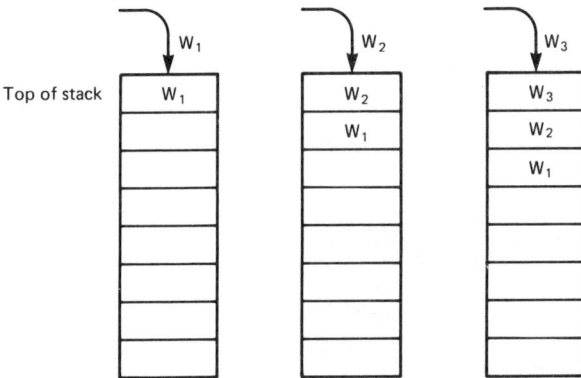

Figure 6.11 Entering words onto a stack

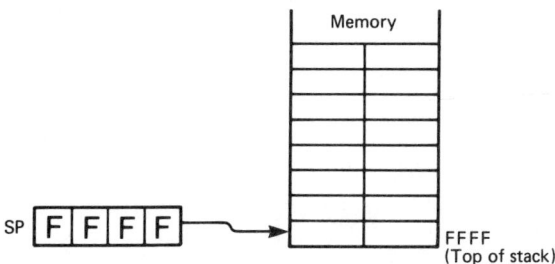

Figure 6.12 The use of RWM for stack operations

When a section of RWM is reserved for stack operation then the information stored is accessed by a 16-bit counter called the stack pointer. One of the first instructions to be written in any program will establish the address of the top of the stack in the stack pointer, as illustrated in Figure 6.12. The loading of the stack pointer can be programmed by writing the load immediate instruction:

LXI   SP,   FFFFH

When data is placed on the stack, the stack pointer is decremented and consequently moves into address areas associated with memory locations whose addresses are <FFFFH. For example, when the contents of the program counter are transferred to the stack at the start of a subroutine, then the most significant byte of the program counter contents will be stored in memory location FFFF−1, and the least significant byte in FFFF−2. This second location now becomes the top of the stack; in other words, the top of the stack moves up and down in sympathy with the contents of the stack pointer.

There are a number of instructions associated with stack operations, and these are shown tabulated in Figure 6.13(a). A block schematic of the 8085A registers is shown in Figure 6.13(b), and those shown unshaded are available for use in stack operations.

Source code	Object code	RTL statement
PUSH B PUSH D PUSH	C5 D5 E5	} ((SP) −1) ← (rh) ((SP) −2) ← (rl) (SP) ← (SP) −2
POP B POP D POP H	C1 D1 E1	} (rl) ← ((SP)) (rh) ← ((SP) +1) (SP) ← (SP) +2
PUSH	PSW	} ((SP) −1) ← (A) ((SP) −2) ← (PSW) (SP) ← (SP) −2
POP	PSW	} (PSW) ← ((SP)) (A) ← ((SP) +1) (SP) ← (SP) +2
LXI SP,addr	31, byte, byte	} (SPH) ← (byte 3) (SPL) ← (byte 2)
SPHL	F9	(SP) ← (H)(L)
XTHL	E3	} (L) ↔ ((SP)) (H) ↔ ((SP) +1)

(a)

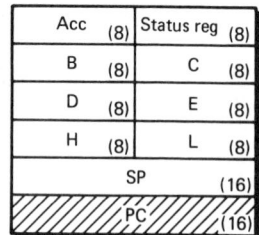

(b)

Figure 6.13 (a) Instructions associated with stack operations (b) Block schematic of the 8085A registers; those shown unshaded can be used in stack operations

The PUSH and POP instructions are used for placing data onto and removing data from the stack, respectively. Each of these instructions is associated with a 16-bit register pair. For example, PUSH B, when executed, places the contents of register pair BC on the stack, while PUSH PSW places the contents of the accumulator and the status register on the stack. The POP H instruction, when executed, results in the return of previously stored data from the stack to the H and L registers.

In line with normal microprocessor practice, pushing data onto the stack from a register pair does not destroy the contents of that register pair. However, the register pair can now be regarded as empty, and the information in them can be overwritten since their contents are now safely stored on the stack. Similarly, popping data from the stack does not destroy the data on the stack, but the memory locations from which data has now been popped can be regarded as empty and can be overwritten with fresh data during a subsequent push operation.

When a subroutine is called it may be necessary to store the contents of all registers as well as the contents of the program counter. The program counter contents are automatically stored when the CALL instruction is executed, but if the contents of the other general purpose registers are to be stored on the stack the first group of instructions in the subroutine will be a series of PUSH instructions:

```
PUSH B  ;  ⎫
PUSH D  ;  ⎬  Store contents of B, D and H
PUSH H  ;  ⎭  registers on the stack
```

Because the stack is a last-in, first-out structure, when the end of the subroutine is reached and the contents of the general purpose registers have to be restored a series of POP instructions written in the reverse order is required:

```
POP H   ;
POP D   ;
POP B   ;
```

Stack instructions can also be used to gain access to the contents of the status register. When the instruction PUSH PSW is executed, the contents of the status register and accumulator are transferred to the stack. A POP instruction will now move this data from the stack to the selected register pair, and if this is followed by the appropriate move instruction, the contents of the status register appear in the accumulator from where they could be moved to a selected output port. The program listing for this operation is:

PUSH PSW	; Acc. and Status reg. to stack
POP H	; Acc. and Status reg. to H and L regs.
MOV A,L	; Transfer contents of status reg. to acc.
OUT 01H	; Output contents of status register

It is a simple matter to write a short program, using the stack instructions, to exchange the contents of a pair of register pairs, and the reader is invited to do this as an exercise.

The LXI instruction listed in the tabulation (Figure 6.13(a)) is used for setting up the stack pointer. This will be one of the first instructions in any program sequence that utilizes the stack.

Besides loading the stack pointer with the LXI instruction, it can also be loaded by the SPHL instruction. This is a 16-bit register transfer function that loads the contents of the H and L registers into the stack pointer. The starting address of the area of memory to be employed in stack operations can be computed by the program, transferred to the H and L registers and then loaded into the stack pointer. The XTHL instruction, on the other hand, changes the data stored at the top of the stack. Data stored in the H and L registers is exchanged with the data stored in the top two stack locations.

6.9 Subroutines

When a program is written, a common set of instructions within that program may appear a number of times; for example, a multiplication routine may be required on a number of separate occasions within the framework of the main program. It would be an advantage to program the common set of instructions once only, and call for them when required. A common set of instructions such as this, forming a subset of the main program sequence, is called a *subroutine*.

If a subroutine is required, a branch from the main program has to be initiated and, at the end of the subroutine, a return has to be made to the main program sequence at the original point of exit. In the 8085A, a branch is initiated by writing the CALL instruction, and a return to the main program sequence is achieved by writing the RET instruction.

The CALL instruction consists of three bytes. Byte 1 contains the instruction opcode while bytes 2 and 3 contain the starting address of the subroutine. This address has to be transferred to the program counter, while the address of the next instruction in the main program sequence is stored in that part of the read/write memory assigned to the stack. These operations are defined by the following RTL statements which appear in the instruction set for the 8085A for the unconditional CALL instruction:

$$((SP) - 1) \leftarrow (PCH)$$
$$((SP) - 2) \leftarrow (PCL)$$
$$(SP) \quad \leftarrow (SP) - 2$$
$$(PC) \quad \leftarrow (byte\ 3)\ (byte\ 2)$$

The interpretation of these RTL statements is that the most significant eight bits of the address of the next instruction in the main program sequence are transferred to the stack location whose address is one less than the present contents of the stack pointer. The least significant eight bits of the address of the next instruction are transferred to the stack location whose address is two less than the present contents of the stack pointer. The stack pointer is decremented by two, and the contents of bytes 2 and 3 of the CALL instruction are transferred to the program counter.

The RET instruction consists of one byte only and that is of course the instruction opcode. When this instruction is executed the address of the next instruction in the main program sequence is returned from the stack to the program counter and the stack pointer is incremented twice. These operations are summarized by the following RTL statements:

$$(PCL) \quad \leftarrow ((SP))$$
$$(PCH) \quad \leftarrow ((SP) + 1)$$
$$(SP) \quad \leftarrow \quad (SP) + 2$$

Entry into and exit from a subroutine is illustrated in the diagram shown in Figure 6.14.

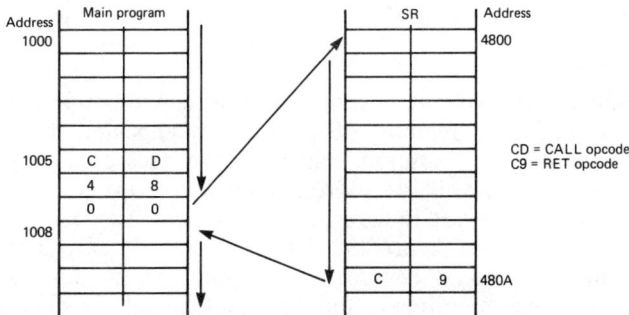

Figure 6.14 Transfer of program control to a subroutine

Subroutines can also call up other subroutines, which leads to the *nesting* of subroutines. This is possible because of the LIFO structure of the stack. Figure 6.15 illustrates the transfer of program control to two subroutines when there is one level of nesting. The main program calls up subroutine 1 and stores the address of the next instruction in the main sequence on the stack. Subroutine 1 at the appropriate point in its program sequence calls up subroutine 2 and the address of the next instruction in subroutine 1 is stored on the stack. At the end of subroutine 2 the RET instruction is written; when it is executed the next instruction in subroutine 1 is returned from the stack to the program counter. When the remaining instructions in subroutine 1 have been executed the terminating RET instruction transfers the next instruction in the main sequence back to the program counter and the main program continues.

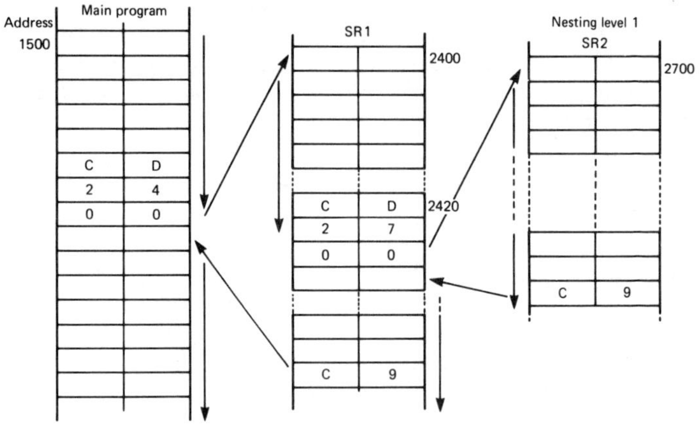

Figure 6.15 The nesting of subroutines

The number of levels of nesting available depends upon the amount of read/write memory that has been allocated by the programmer for stack operations. Normally the stack pointer will be set to the highest available address in read/write memory, and as more data is transferred to the stack it moves to a lower memory address. Clearly, if stack pointer decrementation is carried far enough, the stack could encroach on an area of memory not allocated to stack operations, and this could result in data in memory being overwritten by stack data.

In the examples shown (Figures 6.14 and 6.15) it has been assumed that when a subroutine is called only the address of the next instruction in the main program sequence has to be stored. However, at the point of entry into the subroutine the general purpose registers in the microprocessor may hold data which will be required when the main program sequence is rejoined. In such circumstances the contents of these registers will also have to be transferred to the stack and that will require the utilization of the PUSH instruction described earlier in this chapter. At the end of the subroutine the contents of these registers have to be restored to their original condition, and this can be achieved by transferring data from the stack to the appropriate register by the POP instruction.

6.10 The call and return instructions

In the previous section it was assumed that the entry to and exit from a subroutine was unconditional and the CALL and RET instructions employed were instructions to which no condition was attached. Besides these unconditional instructions there are a number of CALL and RET instructions that have conditions attached to them. As in the case of the conditional jumps described earlier in this chapter, the conditions attached to a call instruction are examined during the execution of that instruction to see whether it has been satisfied. If the required condition has been met program control is transferred to the subroutine and if it has not been met the next instruction following the CALL instruction in the main program sequence is executed.

The conditions applied to call instructions are identical to those applied to jump instructions, and depend upon:

(a) whether the contents of the accumulator are zero, or
(b) the sign of its contents, or
(c) the parity of its contents, or
(d) whether a carry has been produced during the last operation performed by the ALU.

The execution time of this group of instructions depends on whether the attached condition has been satisfied and the subroutine is called, or whether the program counter continues down the main program sequence. If the subroutine is not called, the instruction takes nine T-states or clock periods to execute. If, however, the subroutine is called, a total of 18 T-states is required for instruction execution. The additional time is required to store the program counter contents on the stack and insert the starting address of the subroutine in their place.

Identical conditions are attached to RET instructions. If the condition attached to the instruction is met then program control is transferred back to the calling program at the instruction immediately after the call instruction, but if the condition is not met, the next sequential instruction after the RET instruction is executed. This implies that a return instruction is not necessarily the last instruction in a subroutine; nevertheless, it has to be recognized that at the point of termination of a subroutine a RET instruction has to be written.

Source code	Object code	Condition	Flag condition	RTL statement
CALL, addr	CD, byte, byte	None	None	((SP) −1) ← (PCH) ((SP) −2) ← (PCL) (SP) ← (SP) −2 (PC) ← (byte 3)(byte 2)
CC, addr	DC, byte, byte	Carry	C = 1	
CNC, addr	D4, byte, byte	No carry	C = 0	((SP) −1) ← (PCH)
CM, addr	FC, byte, byte	Minus	S = 1	((SP) −2) ← (PCL)
CP, addr	F4, byte, byte	Plus	S = 0	(SP) ← (SP) −2
CPE, addr	EC, byte, byte	Even parity	P = 1	(PC) ← (byte 3)(byte 2)
CPO, addr	E4, byte, byte	Odd parity	P = 0	if specified
CZ, addr	CC, byte, byte	Zero	Z = 1	condition is met
CNZ, addr	C4, byte, byte	Not zero	Z = 0	
RET	C9	None	None	(PCL) ← ((SP)) (PCH) ← ((SP) +1) (SP) ← (SP) +2
RC	D8	Carry	C = 1	
RNC	D0	No carry	C = 0	(PCL) ← ((SP))
RM	F8	Minus	S = 1	(PCH) ← ((SP) +1)
RP	F0	Plus	S = 0	(SP) ← (SP) +2
RPE	E8	Even parity	P = 1	if specified
RPO	E0	Odd parity	P = 0	condition is met
RZ	C8	Zero	Z = 1	
RNZ	C0	Not zero	Z = 0	

Figure 6.16 The CALL and RETURN instructions for the 8085A

Conditional return instructions, like conditional call instructions, have a variable execution time depending upon whether the condition attached to the instruction is satisfied. If the condition is not satisfied, nine T-states are required for instruction execution, and if it is satisfied 18 T-states are needed.

The tabulation shown in Figure 6.16 gives a list of all the available call and return instructions associated with the 8085A.

Problems

6.1 Write an assembly language program for calculating the 16-bit sum of a series of six 8-bit binary numbers which are stored in six consecutive memory locations.

   After the sum has been formed, and using complement arithmetic, subtract it from FF00H, storing the result in the appropriate number of memory locations.

6.2 Write a program for counting the number of zeros (00H) occurring in a consecutive sequence of 20 memory locations and transfer the number of zeros counted to the output port whose address is 01H.

6.3 Develop a flowchart and write the corresponding assembly language program for

   (a) a subroutine that will determine the number of negative numbers in a block of data stored in consecutive memory locations, and

   (b) a subroutine that will identify and store the address of the most negative number appearing in the same block of data.

   It may be assumed that the length of the block is stored in the memory location immediately preceding that containing the first piece of data and that all data in the block is expressed in 2's complement form. When returning from these subroutines, the number of negative numbers is stored at the bottom of the block and the address of the most negative number is stored in the next memory location.

6.4 Write assembly language programs for subtracting two 16-bit numbers (a) in unsigned form and (b) in signed 2's complement form. For subroutine (a) subtraction instructions may be used, while subroutine (b) should be written without using subtraction instructions.

   The numbers are stored least significant bits first in memory locations 2040H and 2042H. After subtraction, place the result in the H and L registers.

6.5 A block of ASCII characters is stored in successive memory locations whose starting address is in memory location 0100H. A second block of ASCII characters of equal length is similarly stored in successive memory locations whose starting address is 0125H. The length of the two blocks is stored in memory location 0099H. Write a subroutine that compares the two equal length blocks to see if they differ. On returning from the subroutine, any difference in the two blocks should be identified by setting all the 0s in memory location 0150H otherwise all the 1s should be set in this location. Additionally, the address of the first location in the first block that differs from the corresponding location in the second block should be placed in two successive memory locations immediately after the end of the second block.

6.6   Draw a flowchart and write an assembly language program that adds an odd parity digit wherever necessary to each member of a block of ASCII characters. The length of the block of characters is stored in memory location 2000H while the first member of the block is stored in memory location 2041H.

6.7   Draw a flowchart and write an assembly language program for the multiplication of two 8-bit unsigned numbers, placing the product in two adjacent locations in memory.

With the aid of Booth's algorithm, develop a second program which will allow the multiplication of two signed 8-bit numbers.

7 Assembly language programming and software aids

7.1 Introduction

In Chapter 5 it was shown that a program listing could be made in machine code and that in such a listing each byte of an instruction would consist of a pattern of 0s and 1s. The difficulty with machine code programming is that it is prone to error and is tedious to write. Furthermore, an examination of the program reveals little about its nature. Some improvement is gained by converting from a binary listing into a hexadecimal listing. Programming in hexadecimal is less tedious, not quite so prone to error as machine code, but again, a list of hexadecimal numbers does little to improve the understanding of a program to the eyes of a casual observer.

To overcome these difficulties a mnemonic representation is used for instructions and a collection of mnemonically represented instructions is termed an assembly language program. An examination of a program in this form by a casual observer reveals a good deal about the content and nature of the program because of the affinity between the mnemonic form and written English.

In the previous two chapters examples of simple assembly language programs have been given in the text and these have consisted of a series of assembly language statements. There are three different types of statement used in practice. They are:

(1) Instructions of the type described in the previous two chapters. Such instructions are directly executable by the machine for which they have been designed.
(2) *Data descriptors*, which are used to reserve data locations in memory and also for defining constant values.

(3) *Pseudo-instructions* or assembler directives, which provide information that controls the conversion from assembly language into machine code form in a translator or, as it is more commonly called, an *assembler*.

In order to get an accurate translation of assembly language statements they must conform to a rigidly defined set of rules which govern their form and content. This set of rules is referred to as the *syntax* of the language and is analogous to the rules that govern the construction of sentences in written English. Rules are also defined which govern the formation of the three types of statement, previously described, into a program. In this case the rules are referred to as the *program syntax*.

7.2 Source and object programs

It will have been observed in the previous two chapters that in the instruction tabulations the columns containing the mnemonic and hexadecimal forms of the instructions have been headed 'source' and 'object' code respectively. A similar distinction is drawn between an assembly language program and its corresponding machine code program. The former is referred to as the source program, while the latter is called the object program.

7.3 The fields of an assembly language statement

An assembly language statement will define an operation, an addressing mode and an operand address. Additionally, it may contain a comment which is designed to clarify the purpose of the instruction contained in the statement and it may also have a label attached, for reasons which will become clear later. When the statement is to be translated into machine code it must conform with the syntax of the assembler. This varies from assembler to assembler. However, there are some perfectly general rules which apply to all assemblers. For example, the information contained in an assembly language statement is organized in four distinct fields, as illustrated in Figure 7.1.

Figure 7.1 The four fields of an assembly language statement

In practice, the format allows any number of spaces between the fields on a line of code, but the transition from one field to the next is marked by a *delimiter*. A commonly used set of delimiters is tabulated in Figure 7.2. It will also be observed that this table defines the delimiter used between operands in the operand field.

Delimiter	Position
: (space) ; ,	After a label Between operation and operand field Start of comments Between operands

Figure 7.2 Table of commonly used delimiters

The four fields described in Figure 7.1 provide the following information:

(1) *Label field:* this is used to assign a symbolic name to an instruction. Once defined, such a label can appear in the operand field of another instruction in the program listing and is used to refer back to the originally labelled instruction. For example, a label will appear in the operand field of a jump instruction and is used to indicate that a jump is to be made to the instruction where the same label appears in the label field.

(2) *Operation field:* this contains the mnemonic representation of the operation to be performed and is restricted to those mnemonics which appear in the machine instruction set.

(3) *Operand field:* this provides address and data information. An entry in this field can specify a register, a register pair, immediate data, an address, or a label. Address or data can appear in decimal, binary or hexadecimal form and symbols are used to distinguish between the three representations. A decimal representation can be followed by the letter D, hexadecimal by the letter H and binary by the letter B, although it should be noted that other distinguishing marks are frequently used. Memory locations may also be represented by symbolic names, to which values can be attached, and these appear in the operand field in the same way as a label.

Addressing modes may also be identified by an entry in the operand field. For example, the symbol # is sometimes used to indicate the immediate addressing mode and X to indicate indexed addressing. Instructions using the implied addressing mode do not require an operand field so that in this case it is left blank. With direct addressing, the address is contained in the instruction and consequently a distinguishing mark for this addressing mode is not required in the operand field.

In those cases where the operand is an ASCII code character, it may appear in the field in inverted commas or, alternatively, in hexadecimal form. For example, the character 'W' could also appear as 57H.

(4) *Comment field:* this gives a brief description of the purpose of the instruction within the context of the program. There is no limit to the entry in this field except the total number of characters available in a line of program, but if needs be it is always possible to spill over onto the next line.

7.4 Manual assembly

Normally, a program will be assembled automatically by an assembler, but in the case of simple and short program listings assembly can be done manually. The method described here for manual assembly is identical to that utilized by a *two-pass assembler*.

The assembly language program written below will, when executed, determine the maximum number in a block of data. The

Label	Operation	Operand(s)	Comment
MAX:	LXI	H,POINT	; Initialize pointer
	MOV	B,M	; Initialize counter
	XRA	A	; Clear acc. and flags
LOOP:	INX	H	; Increment pointer
	CMP	M	; Compare memory with acc.
	JNC	DECR	; Carry = 0?
	MOV	A,M	; Move memory contents to acc.
DECR:	DCR	B	; Decrement count
	JNZ	LOOP	; Count = 0?
	STA	STORE	; Store contents of acc.
	HLT		; Stop

technique used is to set up a count length equal to the number of elements in the block, clear the accumulator and the carry flag, and compare the first number in the block with the contents of the accumulator. If this number is larger than the contents of the accumulator a carry is generated and the number is then transferred to the accumulator. This procedure is repeated for all numbers in the block, and when the count reaches zero the process of comparison is terminated; the accumulator now contains the largest number in the block. This number is then transferred to a convenient memory location.

The source program given above has been written under the field headings defined in the last section, namely label, operation, operands and comments. Manual assembly of the source program can now be performed in two stages, each of which entails a complete pass through the program.

Stage 1 Determines the memory location into which the opcode or first byte of each instruction has to be loaded, and at the same time creates a label table which associates each label occurring in the source program with a memory address. Assuming that the memory location for the starting address is 2000H then the opcode byte of the first instruction will be allocated to that memory address. This address is regarded as the origin of the program and is placed in the *location counter* whose function is to record the number of instruction bytes in the program during this stage of assembly. Simultaneously, the address for the opcode or first byte of each instruction is recorded in the address column as shown below:

Address	*Object code*	*Label*	*Operation*	*Operand(s)*
2000H		MAX:	LXI	H,POINT
2003H			MOV	B,M
2004H			XRA	A
2005H		LOOP:	INX	H
2006H			CMP	M
2007H			JNC	DECR
200AH			MOV	A,M
200BH		DECR:	DCR	B
200CH			JNZ	LOOP
200FH			STA	STORE
2012H			HLT	

LABEL	TABLE
MAX	2000H
LOOP	2005H
DECR	200BH
POINT	2100H
STORE	2200H

Stage 2 In the second pass through the program listing, the object code column is completed. The mnemonic representation of each instruction is translated into its corresponding hexadecimal form. For instructions that consist of more than a single byte, reference is made to the label table for the hexadecimal representation of the operand contained in the instruction. For example, reference to POINT in the label table gives the operand for the LXI instruction.

At the end of the second stage the assembly is complete, as illustrated in the listing below. This method of program assembly is called two-pass assembly.

Address	Object code	Label	Operation	Operand(s)
2000H	210021H	MAX:	LXI	H,POINT
2003H	46H		MOV	B,M
2004H	AFH		XRA	A
2005H	23H	LOOP:	INX	H
2006H	BEH		CMP	M
2007H	D20B20H		JNC	DECR
200AH	7EH		MOV	A,M
200BH	05H	DECR:	DCR	B
200CH	C20520H		JNZ	LOOP
200FH	320022H		STA	STORE
2012H	76H		HLT	

7.5 Assembler directives

An assembly language program is normally translated by an assembler and *directives* or *pseudo-instructions* are used to control the generation of the object code by the assembler.

The two basic pseudo-instructions that define the start and finish of a program are ORG and END. Three fields are usually associated with assembler directives, namely label, opcode and operand, and in the case of ORG they may be filled as shown below:

Label	*Opcode*	*Operand*
optional:	ORG	expression

The location counter is set to the value of the expression in the operand field, in other words the starting address of the program. In the event of the ORG directive being absent, the address of the first instruction in the program assembly will be 0000H. A program may contain any number of ORG directives and when this occurs it implies that different sections of the program are to be stored in different areas of memory. When the optional label is present, it is assigned the current value of the location counter before it is updated by the expression in the operand field.

The assembler also requires to know where the program ends. It is not always the case that a program terminates with a halt instruction, and for this reason the programmer writes the END directive to inform the assembler of the termination of the program. The three fields of the END directive may be filled as shown below:

Label   Opcode   Operand

optional: END   expression

There can obviously be only one END directive in a source program and clearly it must be the last statement. The expression in the operand field can be used to give the starting address for program execution. If this field is empty then the assembler will assume program execution begins at 0000H.

The location counter acts for the assembler in the same way as the program counter acts for the microprocessor. Its function is to identify where the next instruction in the source program is to be stored in memory. As the assembly language menemonic forms are translated by the assembler, the location counter is incremented. It is essential for the programmer to allocate a memory location to the first instruction in the program, and the assembler will then allocate the succeeding instructions to memory locations in order.

The assembler automatically allocates values to symbols that appear as instruction labels and the value is the current setting of the location counter. For example, in the manually assembled program of Section 7.4 the label MAX: is allocated a value of 2000H and this is entered in the label table. It is possible to define other symbols and allocate values to them using the EQU and SET directives. These symbols are distinguished from the labels dealt with previously in that they are not terminated by a colon.

For the EQU directive, the three fields may be filled as shown below:

Label   Opcode   Operand

Name   EQU   expression

A symbol defined by the EQU directive cannot be redefined during assembly, so it cannot now be used as a label for another instruction. The assembler associates the expression in the operand field with the name, by placing both name and expression in the label table. From that point on, whenever the symbolic name appears in the program it is replaced by the value of the expression in the EQU directive.

An alternative to EQU is the SET directive, which has the following three fields:

Label	Opcode	Operand
Name	SET	expression

The function of SET is identical to EQU except that the name can appear in a number of SET directives in the same program, and consequently the value allocated to the name can be altered throughout the assembly.

The basic advantage to the representation of data constants by symbolic labels is that they all appear at the beginning of the program and a change in a data constant only requires a change in the EQU directive and the program can then be reassembled. Without the use of the EQU directive, the data constant might appear in the operand field of a number of different instructions in the program listing. For a change to be made, each appearance of the constant would have to be located and changed individually.

Another directive available can be used to define a block of memory space. This is the DS (define storage) directive which has the following three fields:

Label	Opcode	Operand
Name	DS	expression

The value of the expression specifies the number of memory bytes to be reserved for storage. Theoretically, the value of the expression can range from 0000H to FFFFH for the case of processor with a 16-bit address bus, but in practice the number of memory locations reserved with this directive must always leave sufficient memory space for the program being assembled. If the expression is zero, no memory space is reserved. When a label is associated with the directive, it is allocated the current value of the location counter. The DS directive then reserves the required memory space by incrementing the location counter by the value of the expression in the operand field.

Two other pseudo-instructions are DB (define byte) and DW (define word). The three fields of DB are:

Label	Opcode	Operands
Optional	DB	expression list

This directive is used to define data constants that are represented by 8-bit words, and it stores each 8-bit value at an address defined by the current setting of the location counter. Individual items in the list are separated by commas and it can contain a number of items. A typical statement defining four one-byte constants is given below

WORD1   DB   0A,42,5F,AB

The optional label WORD1 is assigned the starting value of the location counter and consequently references the first byte stored by this directive.

The other pseudo-instruction has the mnemonic form DW and is used to define 16-bit data constants. The least significant eight bits of the first constant in the expression list are stored at the current setting of the location counter, and the most significant eight bits are stored in the next higher location. This process is repeated until all the data constants are stored.

As in the case of the DB directive, items in the list are separated by commas, and the list can contain a number of items. A statement defining four two-byte constants follows below:

Label   Opcode   Operands

ADDR1   DW   748A,00FF,102A,482C

The label ADDR1 is allocated to the starting address in the location counter and references the first byte of the DW directive.

By utilizing the conditional assembly directives IF, ELSE and ENDIF it is possible to assemble sections of a program providing certain specified conditions have been met. The three conditional assembly directives are described below in terms of the three previously defined fields of pseudo-instructions:

Label   Opcode   Operand

optional: IF   expression
optional: ELSE   –
optional: ENDIF   –

The expression in the operand field of the IF directive is evaluated by the assembler. If the least significant bit is 1, all the instructions between IF and the next ELSE or ENDIF directive are assembled, but if this bit is 0, the assembler omits these instructions from the assembly. These two directives define a block of instructions that may or may not be assembled, depending upon the value of the expression placed in the operand field of the IF directive.

The ELSE directive is optional and may be used to further subdivide the block of instructions between the IF and ENDIF directives, only one of which will be assembled.

Inclusion of conditional assembly directives allows a microcomputing system that is being developed to contain optional features. The system software can include program listings for all these additional features, but, by utilizing the conditional assembly directives, those options that are not required in a particular custom application can be omitted from the assembly process. Similarly, hardware associated with the various options can be developed on separate PCBs which can be either incorporated or omitted from the microcomputer system during production.

7.6 The two-pass assembler

Automatic assembly of an assembly language program is performed by a two-pass assembler. The assembler consists of a look-up table of instruction opcodes which contains an entry for every instruction in the instruction set of the machine. The table contains the mnemonic form of the code, its binary equivalent, information concerning the number of bytes in the instruction, and the nature of the data in the operand. A table look-up procedure is employed to search the table of entries until a match is obtained between the mnemonic form appearing in the operation field of the assembly language statement and the corresponding mnemonic form stored in the assembler. Additionally, the assembler contains the location counter which is used to record the memory locations that have been assigned to the instructions in the assembly language program.

During the first pass, each field contained in the program of assembly language statements is examined sequentially by the assembler and all instructions are allocated memory addresses that are defined by the contents of the location counter. A label table is generated which provides the transformation between all the labels in the program and the value they represent. Additionally, the appropriate instruction opcode is found in the look-up table and extracted, and the location counter is then incremented by the number of bytes associated with the instruction.

The first pass through the program is immediately followed by the second pass. During the second pass, the assembly language statements are examined again sequentially and the corresponding instruction opcodes are located in the look-up table. An examination of the information stored with the opcode reveals whether an operand exists, and the assembler then searches the label table for the binary value of the operand. At the end of the second pass, the object code corresponding to the source code will have been produced in machine code form, and will be stored in the addresses defined by the location counter.

The flowchart describing the tasks performed during the first pass by the assembler is given in Figure 7.3. At the beginning of the pass, the contents of the location counter are initialized to zero. The first line of code is then examined by the assembler. Normally it will be the assembler directive ORG, and will therefore not have a label. At this point the location counter will be set to the value appearing in the operand field of the ORG directive, this being the starting address of the program.

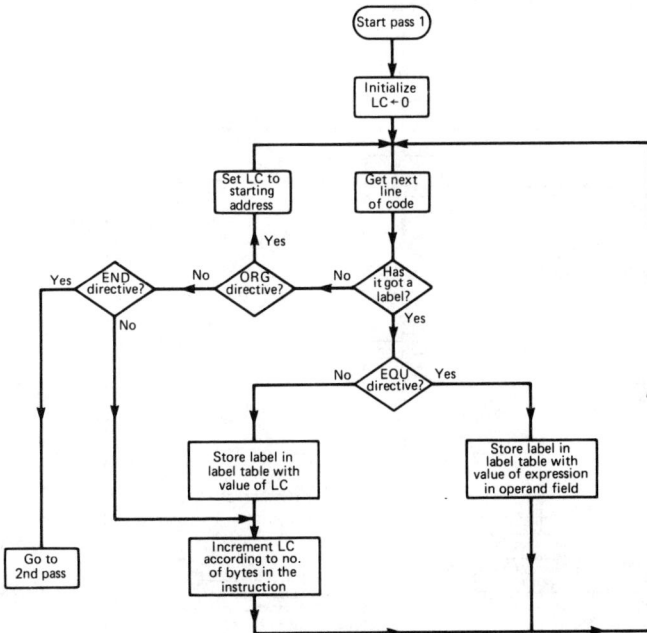

Figure 7.3 Flowchart for first pass of assembler

The next line of code will have a label, and the assembler will need to know whether it is the assembler directive EQU. If it is not, the label will be stored in the label table with the current contents of the location counter; if it is, the label will be stored in the label table in company with the value appearing in the operand field of the EQU directive. The location counter is now incremented by the number of bytes contained in the instruction and the next line of code is examined. If this line of code does not have a label, is not ORG or END, then the location counter is incremented by the number of bytes in the instruction. This process continues until the END directive is recognized, and at that point the first pass is terminated.

When the second pass is entered by the assembler, the assembly language statements are again examined sequentially. A flowchart describing the sequence of events in the second pass is given in Figure 7.4. This flowchart is identical to the one describing the first pass, but now when a line of code is examined the assembler requires to know whether it is a directive. If it is not a directive,

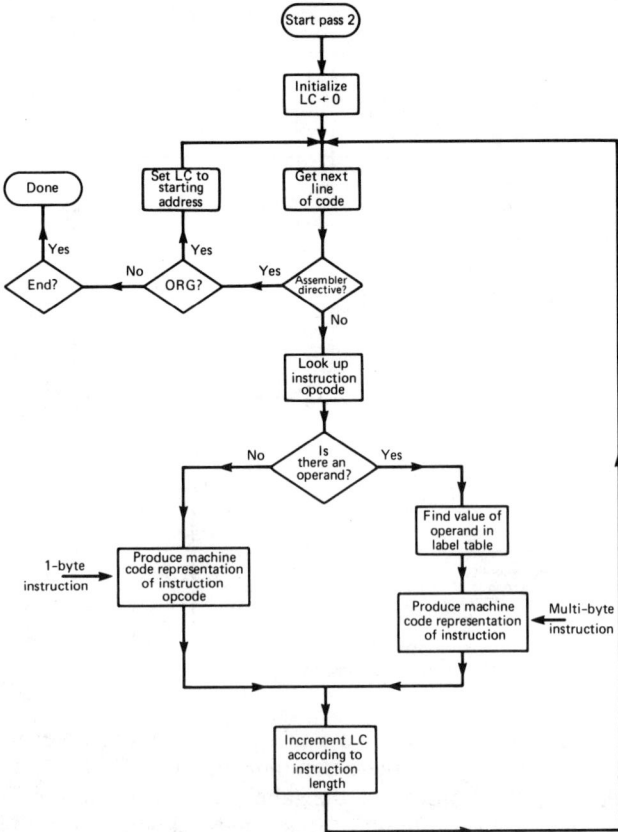

Figure 7.4 Basic flowchart for second pass of assembler

the assembler searches the look-up table until the contents of the table and the assembly language statement match, and then examines the information stored. This will indicate whether there is an operand associated with the instruction opcode.

If there is no operand, the instruction consists of a single byte and the machine code representation of this byte is now associated with the current contents of the location counter, which is, of course, the address the instruction will occupy when loaded into memory. For an instruction consisting of more than one byte there is an operand, and its value is found by searching the label table. The operand and the instruction opcode in binary form are then associated with consecutive contents of the location counter. Finally, the location counter is incremented in accordance with the instruction length.

An important task of an assembler is to check for possible errors in the program. For example, an invalid mnemonic form of an instruction opcode cannot be translated by the assembler since it does not know its equivalent machine code form. The assembler then prints an error message to inform the programmer that there is an error in a specified line of code. The ability of the assembler to check for possible errors is referred to as *error diagnostics*.

7.7 Program execution time

The time required to execute a program is obtained from a knowledge of the number of states n required during program execution. Knowing the periodic time of the clock t_c the number of states can be translated into execution time t_e by the use of the equation

$$t_e = nt_c$$

If a program does not contain any jump instructions, determination of the execution time is straightforward and simply calls for a tabulation of the number of states required for the execution of each instruction as specified by the instruction set. The tabulation is then summed to give the overall execution time. If, however, there are jumps in the program listing which generate program loops then a group of instructions is executed during each traversal of the loop, and to compute execution time the number of loop traversals must be known.

The program listing given below implements 8-bit division using a process of repeated subtraction. If, after a subtraction has taken place, a carry is generated then the divisor > dividend; alternatively, if a carry is not generated the divisor < dividend, and in this case a 1 is placed in the quotient register (L). Subtraction takes place eight times, so the loop shown in the flowchart (Figure 7.5) is traversed seven times before the iteration counter contents have been decremented to zero.

States

	LXI	H,0CH	10
	MVI	C,03H	7
	MVI	B,08H	7
TWO:	DAD	H	$8 \times 10 = 80$
	MOV	A,H	$8 \times 4 = 32$
	SUB	C	$8 \times 4 = 32$
	JC	ONE	$8 \times (7 \text{ or } 10) = 56 \text{ to } 80$
	MOV	H,A	$8 \times (0 \text{ or } 4) = 0 \text{ to } 32$
	INR	L	$8 \times (0 \text{ or } 4) = 0 \text{ to } 32$
ONE:	DCR	B	$8 \times 4 = 32$
	JNZ	TWO	$(7 \times 10 + 1 \times 7) = 77$
	MOV	A,L	$= 4$
	STA	0100H	$= 13$
	HLT		$= 5$

355 to 443

The total time taken to execute this program is variable and depends upon the number of 1s that appear in the quotient after the 8-bit division. After every subtraction the carry has to be examined by the JC instruction; if it is 1, the next instruction to be executed is DCR B, and if it is 0, subtraction is possible and the next instruction is MOV H,A.

The minimum time for execution occurs when the quotient Q = 00H. Then

$$t_{min} = 355 \times t_c$$

and for a 3 MHz clock

$$t_{min} = 118 \, \mu s$$

The maximum time for execution occurs when the quotient Q = FFH. Then

$$t_{max} = 443 \times t_c$$
$$= 148 \, \mu s$$

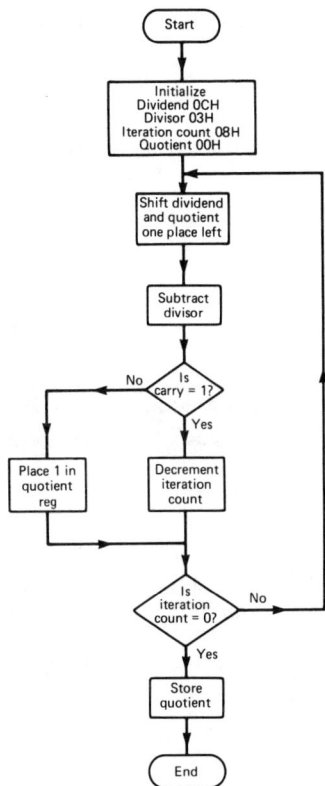

Figure 7.5 Flowchart for 8-bit division

7.8 Software generation of time delays

A microprocessor is frequently used in a control capacity and as such may be required to switch an external circuit at specified time intervals. The time intervals between switching have to be generated by software in the processor.

A software delay can be generated by the program listing given below:

States

```
        MVI   B,50D     7
ONE:    DCR   B         4
        JNZ   ONE       10 (for a jump)
                         7 (for no jump)
```

The delay loop in this program consists of the two instructions, decrement and jump, and the number of traversals of this loop is specified by the count set up in register B. Execution of the program requires 49 traversals of the loop, each traversal consisting of 14 states, hence the execution time is given by

$$t_e = [(49 \times 14) + 18]t_c$$

and for a 3 MHz clock

$$t_c = \tfrac{1}{3} \times 10^{-6}\mu s$$

Hence

$$t_e = 2356\,\mu s$$

The maximum delay that can be generated occurs when the count set up in register B is FFH or 255D. Thus the maximum delay time is

$$t_e = [(255 \times 14) + 18] \times \tfrac{1}{3}\,\mu s$$
$$= 1196\,\mu s$$

Frequently, the required delay is specified, and the programmer has to calculate the required count. For example, for a delay of 500 μs

$$500 = [(Cnt - 1)14 + 18]\tfrac{1}{3}$$
$$Cnt = 107D$$
$$= 6BH$$

7.9 Generation of a repetitive waveform using software delays

It is required to generate the repetitive waveform shown in Figure 7.6(a). This can be achieved by setting and resetting an SR flip-flop, using the circuit arrangement shown in Figure 7.6(b).

The second byte of the output instruction gives the address of the output device selected. In this example, if the address is F8H then the 3-to-8 line decoder is selected and, additionally, the output 0 of this device goes low, and is then inverted to give a logical 1 on the set input of the flip-flop. If, on the other hand, the address provided by the output instruction is F9H, output 1 of the decoder goes low, is then inverted, and provides a logical 1 for the reset line of the flip-flop. The program listing for the required waveform is given below, and it is left to the reader to analyse the program and check the counts placed in the B and C registers.

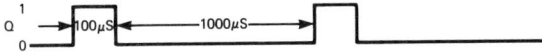

(a)

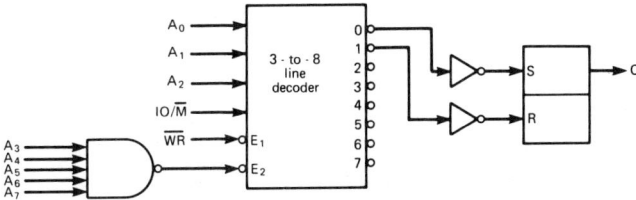

(b)

Figure 7.6 Generation of a repetitive waveform using a microprocessor (a)
Required waveform (b) Circuit configuration for its generation

			States
THREE:	MVI	B,15H	7
	MVI	C,D4H	7
	OUT	F8H	10
ONE:	DCR	B	4
	JNZ	ONE	7 or 10
	OUT	F9H	10
TWO:	DCR	C	4
	JNZ	TWO	7 or 10
	JMP	THREE	10

7.10 Subroutine linkage

The transference of data from a calling program to a subroutine,
and vice versa, is a problem created by the use of subroutines. The
data transferred is usually called a *parameter* and the transference
of parameters to a subroutine is termed *parameter passing*.

There are a variety of methods used for subroutine linkage, and
the parameters may be passed via

(a) the internal registers of the machine,
(b) specially reserved locations in memory,
(c) memory locations identified by a pointer,
(d) the stack.

When the number of pieces of data to be transferred is less than the number of internal registers available in the machine, it is convenient to provide subroutine linkage via these registers. The data to be passed is programmed into them, and the subroutine is then called by the execution of the CALL instruction. The data is processed by the subroutine and the result is placed in a predetermined register, or some location in memory, prior to writing the RET instruction.

Since there is no division instruction available in the instruction set of the 8085A, a subroutine can be written to perform this operation by, say, the restoring method. The subroutine will assume that the calling program has set up the parameters in the internal registers of the machine. For this example the parameters are the divisor, the dividend, and the iteration count. After the division has been executed the quotient appears in a predetermined register and from there can be transferred to memory. Since the contents of the H and L registers can be stored direct, it would be convenient to use this pair of registers for temporary storage of the quotient.

Alternatively, parameters and results can be transferred from the calling program to the subroutine via memory locations reserved in RWM. For example, in the case of the division subroutine previously described, the divisor, the dividend and the iteration count could have been sited at reserved locations in memory; furthermore, the quotient can be returned to the calling program via another location reserved in memory.

Earlier in this chapter it was described how reserved memory locations can be established by the use of the define storage, DS, assembler directive. The operand field of DS can be used to identify those memory locations to be reserved for the parameters which have to be passed.

When a subroutine requires a large number of variables they can be established in consecutive memory locations, and access to this stored data can be provided by the contents of those internal registers used as pointers. As an example of this technique of subroutine linkage, the program listings given below provide for the addition of two multiple-word decimal numbers.

The first program listing is the calling sequence and sets up the pointers in the HL, DE and BC registers respectively, places the number of bytes to be added in the accumulator, and defines the top of the stack. The second listing constitutes the subroutine, and this is the software implementation of the algorithm that provides the rules for decimal addition. It also arranges for the return of the result via the memory locations initially defined by the contents of registers BC.

	List 1		*Calling Sequence*
	LXI	SP,2000H	; Load stack pointer
	LXI	H,0800H	; Set pointer to address byte 1
	LXI	D,0805H	; Set pointer to address byte 2
	LXI	B,080AH	; Load no. of bytes into acc.
	LDA	0900H	
	CALL	DECAD	
	HLT		

	List 2		*Decimal addition SR*
DECAD:	PUSH	B	; Place result address on stack
	MOV	B,A	; transfer bytes to be added to Reg. B
	XRA		; Clear acc. and carry flag
LOOP:	LDAX	D	; Load acc. with byte addressed by DE
	ADC	M	; Add acc. to contents of memory loc. addressed by HL
	DAA		; Convert contents of acc. to BCD
	XTHL		; Place result address in HL
	MOV	M,A	; Move acc. to memory location addressed by HL
	INX	H	; Increment result address
	XTHL		; Transfer result address to stack
	INX	D	; Increment D
	INX	H	; Increment H
	DCR	B	; Decrement B
	JNZ	LOOP	; Jump if count is not zero
	POP	B	; Return result address to B
	RET		; Return to calling program

An examination of the subroutine reveals that in this case the stack also has to be used as temporary storage for the result address, in order that the byte count can be placed in register B. Later in the subroutine, after addition has taken place, the contents of the HL register has to be exchanged with the result address previously stored on the stack so that the result of the addition can be stored in RWM. Immediately after storage, the result address is incremented and is transferred back to the stack using the XTHL instruction.

It follows from the comments made in the preceding paragraph that the stack can also be used to pass parameters that have been placed there by the calling sequence to the subroutine. During the execution of the CALL instruction, the return address is pushed onto the stack. The parameters and the return address form a structure called the *stack frame*. To obtain the parameters from the stack, the return address must be released by the POP instruction and saved, for convenience, in one of the internal registers of the machine or, alternatively, in reserved memory locations. The parameters are now unblocked and can be released from the stack as needed. When all the parameters have been processed, the return address is pushed back onto the stack again, and the RET instruction can be executed, thus restoring control to the calling program at the point of the return address.

If, by any chance, all the parameters have not been used by the subroutine, the base of the stack will have been shifted to a memory location having a lower value than its initial value. Assuming this process is repeated every time the subroutine is called, the base of the stack will move down progressively. If this takes place a sufficient number of times the base of the stack will encroach on memory space that has been allocated for other purposes.

7.11 Re-entrancy

Complex situations can arise when a machine is engaged in the concurrent handling of several tasks. The ability to share a subroutine among a number of tasks is called *re-entrancy*, and this type of subroutine is called a *re-entrant subroutine*. Such a situation can arise when a machine performs one task by executing the main program. However, this program can be interrupted by any of a number of other processes, and at the point of interrupt the machine enters an interrupt service routine which performs a second task and may require to call the same subroutine currently being executed by the main program.

The sharing of a re-entrant subroutine in this way is illustrated in Figure 7.7. The main program performing task X requires the processing provided by the re-entrant subroutine SR. When an interrupt occurs, task X temporarily relinquishes the control of the subroutine before the required processing has been completed. On completion of task Y the main program regains control of the re-entrant subroutine and recommences processing where it left off at the point of interrupt.

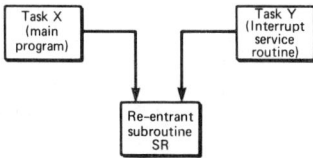

Figure 7.7 Sharing a re-entrant subroutine

Since the re-entrant subroutine was processing data at the point of interruption, arrangements have to be made to ensure that data is not destroyed. In a situation such as this, the benefit of stack operations are self-evident. When an interrupt occurs the contents of all internal registers and the status flags are pushed onto the stack and are not subjected to any manipulations during the interrupt service routine. At the end of the routine, registers and flags are restored to their original condition, and the interrupted program can continue to be executed.

7.12 Macros

During program development a programmer may notice that a specific sequence of assembly language instructions occurs on a number of different occasions. However, on each repetition of the sequence the parameters associated with it may be different. It would be advantageous to the programmer if the original sequence could be generated again with a new set of parameters, without having to rewrite the source code. The software device used for this purpose is called a *macro*, which is, in effect a pseudo-instruction that defines a group of assembly language instructions. Use of macros reduces programming time, tedious repetition, and the probability of error. Additionally, a library of macros can be developed which can be used by any programmer, thus avoiding duplication of effort.

At first sight, macros and subroutines appear to be substantially the same. Both assist program structuring and avoid repeated development of frequently used routines. When a subroutine is called by the main program a jump occurs and there is a discontinuity in the program address sequence. Macros, on the other hand, when called, simply generate a sequence of assembly language instructions which are inserted in the main program sequence. If, for example, the macro was called six times the corresponding sequence of assembly language instructions would appear in six places in the main program sequence. To distinguish between a macro and a subroutine, it is common practice to refer to the macro as an *open subroutine* and a normal subroutine as a *closed subroutine*.

Perhaps the most significant difference between macros and subroutines is that a program can call only a single version of a given subroutine. A macro, on the other hand, does not generate the same source code each time it is called. Simply by changing the parameters in a macro call, the generated source code may be changed.

When an assembly language has a macro-facility, a *macro-assembler* is required to translate it into a normal assembly language program. Conversion from a macro-based language to executable machine code is illustrated in Figure 7.8. The

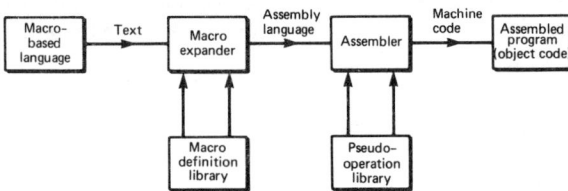

Figure 7.8 Macro-based language translation

macro-based source program is read into the *macro-expander* which, in conjunction with a *macro definition library*, generates the assembly language program. In effect, the macro-expander may be regarded as the front end of the assembler program which, in conjunction with the pseudo-instructions library, translates the output of the macro-expander into executable machine code. Since each macro call has to be expanded before assembly, a macro-based program may well have a long translation time. In effect, there are two conversions to be implemented before the object code is produced.

The programmer, when employing macros, allocates a symbolic name which represents a sequence of assembly language instructions. For every occurrence of the macro in the main program sequence, the macro-expander substitutes the defined sequence of instructions.

The format of a macro is:

Label	Code	Operand
name	MACRO	parameter
	(macro body)	list
	END M	

The name is the symbolic name allocated to the macro and referred to previously, while the macro body is the sequence of assembly language instructions which replaces the macro label when macro expansion takes place. In the operand section of the format, the parameter list appears.

A typical sequence of instructions appearing in many programs is

```
LXI   addr.
MOV   r,M
```

The LXI instruction loads a memory address into the H and L registers, while the MOV instruction transfers the contents of the address pointed to by the H and L registers into register r. The corresponding macro to this sequence is allocated the label INDLO, and is written

```
INDLO   MACRO   Reg, Addr.
        LXI     Addr.
        MOV     Reg, M
        END M
```

The parameters in this case would be the register to which data has to be moved, and the address in memory where the data is to be found.

7.13 High level languages

In this chapter, emphasis has been laid on assembly language programming. An assembly language is described as a low level language since it is only one stage removed from machine code. To convert an assembly language statement to machine code, an assembler is required. An alternative programming technique is to use a high level language such as BASIC, FORTRAN or PASCAL, which offer a number of basic advantages when compared with assembly language programming.

The most striking advantage of high level programming is revealed by writing the same program in (a) high level language

(b) assembly language and (c) machine code. The example chosen
for this comparison is the addition of two numbers:

High level language	Assembly language	Hex	Machine code
LET SUM =	LDA	3A	0011 1010
NUM1 + NUM2		00	addr. low byte
		20	addr. high byte
	MOV B,A	47	0100 0111
	LDA	3A	0011 1010
		01	addr. low byte
		20	addr. high byte
	ADD B	80	1000 0000
	STA	32	0011 0010
		02	addr. low byte
		20	addr. high byte

To program the addition operation, one high level statement, or
five assembly language statements, or eleven lines of machine
code are required. Clearly, assembly language programs will take
longer to write and are more prone to error. Furthermore, an
assembly language program is difficult to understand unless the
reader is familiar with the microprocessor being used, while a high
level version of the program is short and to the point, and easily
understood.

Because an assembly language is associated with a particular
processor, and since compatibility of software is not usual, the
programmer, needs to learn a new assembly language whenever a
new processor is selected. This is not the case with a high level
language, which will run on any machine given the right kind of
software support.

Before a high level language can be used in conjunction with a
particular machine, the high level language statements have to be
translated into machine code. The software device that performs
this operation is called a *compiler*, and the translation process
from high level language to machine code is described in Figure
7.9.

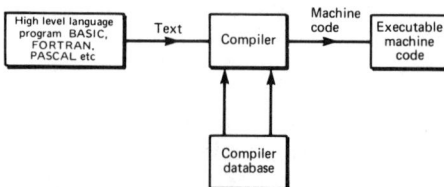

Figure 7.9 High level language translation

7.14 Compilers

A compiler is used to translate a high level language program into executable machine code. A high level language is machine independent, and a program written in such a language can be run on a microprocessor whose library contains a compiler for that language. Because a high level language is more remote from executable machine code than an assembly language, a compiler is more complex than an assembler and requires considerably more development time.

In practice, a compiler is a somewhat rigid translator, and the binary code generated when translating a given algorithm is likely to be more lengthy than that generated by direct assembly language programming. This implies that the machine code program generated by the compiler will require more storage space and will take longer to execute than the corresponding well-designed assembly language program. On the other hand, efficient assembly language programming depends upon the proficiency and skill of the programmer, and this is clearly a variable quantity.

Some compilers translate into assembly language, and executable machine code is then produced by an assembler. For both types of compiler, a listing of the translation is produced and syntax errors are detected and indicated.

If the translated program is not required immediately it is useful to *dump* it from the working storage space into more permanent storage such as a disk. Sequences of instructions that occur repeatedly, and appear in a number of programs, are also stored in a file on disk. A collection of such files is referred to as a source library. When a particular sequence is required at a specified point in the program a statement such as INCLUDE COPYFILE is written, and this will identify the required file and include its contents in the program sequence.

7.15 Interpreters

Both compilers and *interpreters* are translators, but they use quite different techniques. A compiler translates the whole of the high level source program before execution takes place. This is analogous to a letter in English sent to a recipient in a foreign country who does not understand the English language. Before the letter can be understood and acted upon it has to be wholly translated into the recipient's language. An interpreter, on the other hand, translates a single high level statement and executes the corresponding sequence of instructions now appearing in executable machine code. This situation is analogous to the function of an interpreter at an international conference. Progress is slow when this technique is employed because a pause will always occur before the interpreter begins to translate. In the computing world, program execution will be ten to twenty times slower when an interpreter is used.

The action of the interpreter is to translate one high level statement at a time, expand it into one or a sequence of machine code instructions, which are then executed on the processor. When that instruction sequence has been executed, the interpreter fetches the next high level source statement and the procedure is repeated. The process may be described as one of 'translate and execute' for each high level statement fetched. The most commonly used language with the interpretive technique is BASIC.

In a scheme employing an interpreter, the interpreter can operate directly on the high level language statements provided by the programmer. For this technique, only a source file is required. By comparison a compiler translation technique requires two files, one for executable machine code, and also a source file which is needed if modifications have to be made to the program. Both the source and execution file have in this case to be saved, and a section of disk will be reserved for them. For the interpreter the source file is the one which is run when the program is executed, and it is also the one which is modified if program changes are required.

7.16 Monitors and operating systems

The minimum aids required for any computing system are a translater and a *monitor*. For example, the SDK 85 kit referred to in an earlier chapter has a hexadecimal input through a keyboard which is then translated into machine code. A primitive system of this nature can be greatly improved by the provision of more sophisticated aids such as assemblers, compilers and interpreters.

A *monitor*, or *supervisory program*, is in effect an elementary *operating system*. It controls the operation of the computer, albeit at a very elementary level. Facilities are provided for loading programs through a keyboard, examining and changing the contents of a selected memory location, examining the contents of machine registers, single-step operation, etc. The monitor program is normally resident in ROM and its facilities are immediately available on power-up. In more complicated computing systems an enhanced monitor is employed and is usually referred to as an operating system.

Such a system consists of a suite of programs that control the operation of a computing system so that optimum performance is achieved. A basic function of an operating system is to provide the computer user with easy access to, and control of, the hardware of the system. The most common operating system in use with small computers at the present time is CP/M (Control Program/ Monitor).

It is common practice to have the operating system stored on disk and the initial problem is how to get it into machine memory. This problem is overcome by the *bootstrap program* which is loaded into memory when the computer is first switched on, and remains there until it is switched off. Execution of the bootstrap program allows the *loader program* to be transferred to memory. The function of the loader is to read machine code programs into memory, so it is now possible to transfer the rest of the suite of programs that comprise the operating system from disk to memory.

When the operating system has been loaded, it monitors the machine keyboard while awaiting a command signal from the operator. On the receipt of such a signal, the operating system responds. Command signals order the machine to perform those tasks specified by the operator, such as initiate the execution of an applications program. When the initiate signal is received, the application program is transferred to memory and control passes to this program, which is now executed. At the termination of the program, control is returned to the operating system and the machine awaits a further command from the keyboard, which is now being monitored by the operating system again.

The operating system is designed to improve the computing efficiency of the machine, which is measured in terms of *throughput*. This is defined in terms of the quantity of processing completed in a specified time interval and is governed by the speed of the electronic circuitry used in the machine and the facilities provided by the operating system.

An important component of the operating system is the loader program which, when executed, allows the transference of all other programs, both applications and non-applications types, to memory, and prepares them for execution. There are two types of loader:

(a) *Absolute loader*, providing the simplest possible loading scheme. Machine code programs are stored on punched cards or paper tapes after translation. The function of the loader is to accept these machine code programs and transfer them to memory locations specified by the user.

(b) *Relocating loaders*, which provide the additional facility of arranging programs and subroutines efficiently before transferring them to memory.

An important function of an operating system is to provide a simple and convenient method for storing data. The generally accepted storage medium is either floppy or hard disk, although an alternative technique is to use magnetic tape held in a cassette. A disk needs to be prepared before it can be used for storage and one of the functions of the initialization process is to provide a symbolic label for it which can be read by the operating system. The label will be chosen so that it uniquely identifies the disk.

The technique used for storing data on a disk is a *filing system*, which contains a number of individual files. Each file has a stand-alone identity which consists of a stream of characters of arbitrary length. The file could be an applications program, a notice to be circulated, a quotation, or a set of data.

The operating system should be capable of manipulating and storing files; for example, it should allow the generation and deletion of files, access for reading, writing and modification, and reference to a particular file with the aid of a symbolic label. To fulfil the last function the operating system should provide a *directory* of files which contains the file labels and the disk address. Given the symbolic name of the file, examination of the directory will reveal the disk address.

A number of file operations should be provided by the operating system, such as create file, display file, delete file, copy file, print file, and rename file. Any invalid operations should be identified by the operating system and an error message should be generated and displayed. Operating systems should also provide facilities for dealing with a sequence of commands without intervention from the operator.

Multi-user computer systems are now commonly available where several terminals are simultaneously connected to a central computer. For this type of system a multi-user operating system is required to control the computer and its associated peripherals, particularly shared resources like the CPU and a printer. A system such as this can be operated on a *time sharing* basis, where users at different terminals are allowed a series of spaced short time intervals to perform their tasks. A single memory can be employed. The program currently being executed is stored in memory for a short time period. The program is returned to disk after the time interval and is replaced by the second program whose execution also continues for a short period of time before it is superseded by a third program, and so on. The specified time-slices effectively interrupt each program when it is running, although the individual users do not notice this because an output device such as a printer is slow in comparison with the speed of operation of the computer.

Alternatively, a switch bank of memories may be used, one for each of the user terminals. At any given instant of time the CPU can only execute one program, but it may easily be transferred to a second program by switching the memory bank. This method has the economic disadvantage of requiring a separate memory for each user terminal but this can be traded against the reduction of the program changing overheads required in the single memory technique.

7.17 Program libraries

A library is a collection of useful program sequences which the computer user has built up over a period of time. The function of such a collection is to eliminate the drudgery of repetitive programming of these frequently occurring sequences. Repetitive program sequences of this kind have been encountered in an earlier chapter where they have been referred to as subroutines. The types of library subroutine available are frequently mathematical in nature, and are concerned with tasks such as addition, subtraction, division and multiplication. Additionally, a program library is likely to contain subroutines for square roots, exponential and trigonometrical functions, etc.

Besides a library of subroutines, there will also be a library of utility programs whose function is to perform well-defined tasks that reduce programming labour. A typical example of a utility is a *text editor* whose function is the production of a correct and well-organized text which more often than not will be an assembly program listing. The program development process is illustrated in the block diagram of Figure 7.10. This diagram indicates that the

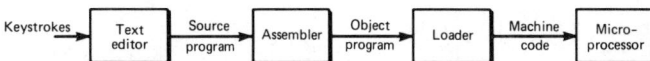

Keystrokes → Text editor → Source program → Assembler → Object program → Loader → Machine code → Micro-processor

Figure 7.10 Program development with software aids

source program is edited with the aid of the keyboard before it is assembled and loaded.

The text editor program facilitates the creation and modification of symbolic source programs and other text materials. By means of commands initiated by the programmer at the keyboard, the editor can create and delete characters, a line, or groups of lines. In most systems the editor is on-line and the response to the commands is immediate.

Originally, editing was carried out on a teletype, which had the disadvantage that corrections could only be made one line at a time. With the advent of VDUs it is now possible to display on the screen the section of a file that has to be edited. This allows the programmer to amend the text visually and return it to the file on disk. Amendments are made by positioning a cursor to the point on the screen where an error has been observed. Characters can be overwritten, inserted or deleted with the aid of control keys on the keyboard.

Problems

7.1   The program given below is used to find the maximum number stored in a block of data. Assemble the program manually, and obtain a label table similar to the one formed by an automatic assembler.

```
        LXI     H,2500H
        MVI     C,19H
        SUB     A
LOOP:   INX     H
        CMP     M
        JNC     CNTD
        MOV     A,M
CNTD:   DCR     C
        JNZ     LOOP
        STA
        HLT
```

What other information would an automatic assembler require? Illustrate your answer with examples.

7.2   Construct a program trace for the program shown above and calculate the execution time of the program, assuming the 8085A clock frequency is 3.125 MHz.

7.3   It is required to produce a pulse at output port 01H of time duration 10 µs with a periodic time of 100 µs. Write an assembly language program to satisfy the given specification.

8 Program controlled input/output data transfers

8.1 Introduction

There are three main techniques presently employed in all computing systems for transferring data between the system and an external device, normally referred to as a peripheral, in both directions. The three methods are:

(1) Program controlled I/O
(2) Interrupt driven I/O
(3) Hardware controlled I/O.

In general, data is transferred in parallel form, a word at a time. However, the advent of computer networks has stimulated the use of serial data transmission systems that provide a facility for error checking. This requires a change of data from parallel to serial form at the transmitting end, and the reverse change at the receiving end.

For program controlled I/O, an I/O operation is initiated when an I/O instruction is encountered in the main program sequence. The transfer takes place via an I/O port to the accumulator of the microprocessor and thence to memory for storage, or to the ALU for processing. Before the data can be transferred it has to be prepared, and when it is ready for transfer a flag or status signal is provided to indicate that this is so. The processor has to examine the status bit regularly in order to detect the availability of data and this requires the additional operation of the transfer of the status bit to the machine. The method is inherently slow, and several instructions have to be executed to transfer a single byte of data. Programmed I/O is a standard feature of a computing system; it has proved to be very flexible but will not provide the high transfer rates of the hardware controlled system. The method can, if necessary, accommodate several peripherals if a software polling system is employed.

In the case of interrupt driven I/O, the data transfer is initiated by the peripheral and not the processor. When the data is ready, the peripheral raises an interrupt signal which is supplied to the processor. This signal is examined once every instruction cycle and, providing the interrupt has not been disabled by software, the processor suspends execution of the main program and enters an interrupt service routine. Control is transferred to this routine and after the processor has received all the data the machine is returned to the control of the main program sequence.

For a hardware controlled I/O, a transfer takes place directly between a peripheral and memory, or vice versa. Transfer by this method is called *direct memory access* (DMA) and is under the control of a DMA controller which is, in essence, a second processor that takes over control of the bus system during the data transfer. The method provides a fast transfer of data and is useful when large blocks of data are to be transferred from, say, a floppy disk to memory.

This chapter will be concerned only with program controlled I/O, while the other two commonly used techniques will be discussed in Chapter 9.

8.2 I/O interfacing

Each peripheral is a unique problem for the system designer, and it is, of course, conceptually different from the microprocessor. Many peripherals are electromechanical devices, while the processor and its associated memory are electronic in nature.

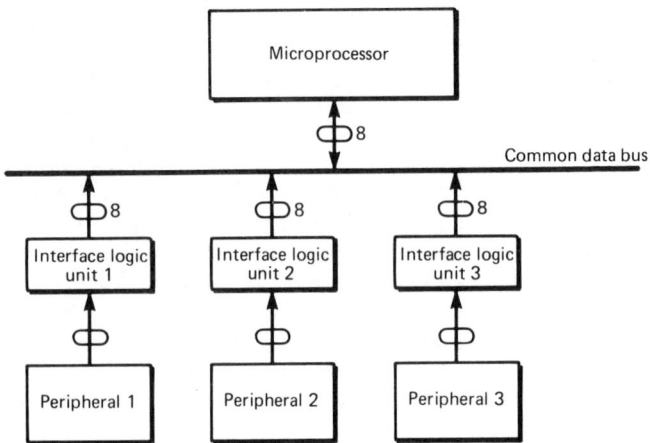

Figure 8.1 Communication link between a microprocessor and a group of peripherals

Peripheral speeds in most cases are different from the speed of the processor, some being slower, others faster, and, additionally, their operation is asynchronous with respect to the clocked operation of the processor. To reconcile these differences, a logical interface unit is placed between the peripheral and the processor, and communication between the two devices is conducted via the common data bus, as illustrated in Figure 8.1.

In its simplest form, the interface logic unit consists of a unidirectional register or input port, whose operation is controlled logically, as shown for an 8-bit system in Figure 8.2. The register has a unique address and, when this is combined with a read signal, originating in the microprocessor, a device select strobe is generated which enables the tri-state output of the register and releases the data onto the common data bus and thence to the accumulator in the processor.

An operation such as the one described can only be implemented providing the processor is aware that the data is available. A data available signal is generated by the peripheral and is stored in a single-bit status register. The microprocessor examines the content of this register at regular intervals and releases its content onto the data bus and thence to the accumulator by means of a second device select strobe generated from a combination of an address and a read signal.

Besides the single port system described above, microprocessor manufacturers are also able to supply more complex I/O chips which function as *programmable peripheral interfaces*. These devices contain a number of I/O ports or registers as well as control and status registers. The I/O ports can be individually programmed to operate in either an input or an output mode. Some manufacturers also provide multi-function devices which, for example, provide storage facilities as well as programmable I/O ports.

8.3 Isolated and memory mapped I/O

There are two distinct ways of manipulating the data that has to be transferred in both directions between processor and external peripheral. These are referred to as the *isolated* I/O mode and the *memory mapped* I/O mode.

In the isolated I/O mode, the microprocessor utilizes the input (IN) and output (OUT) instructions contained in the machine instruction set. These instructions consist of an opcode and a port address. After the opcode has been decoded, a control signal

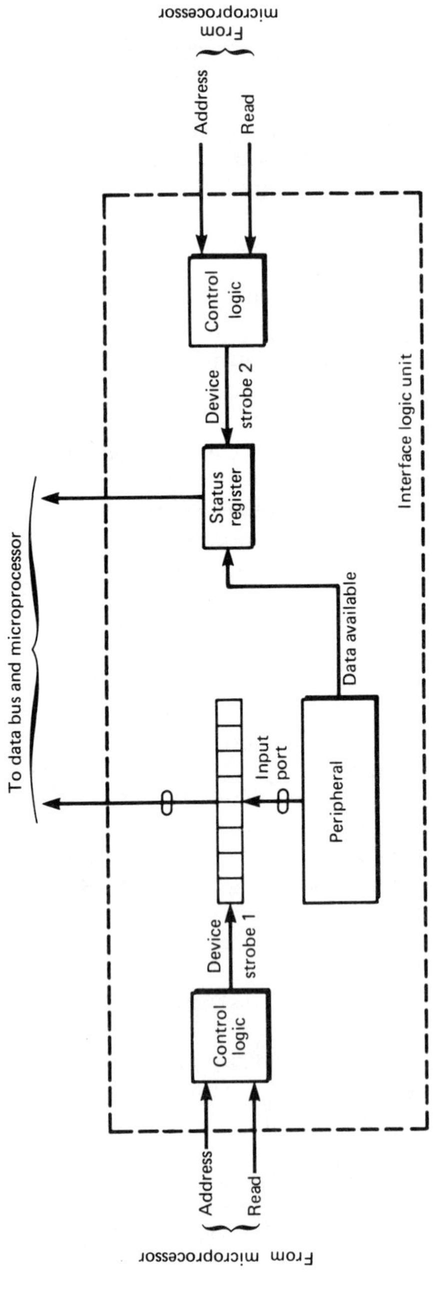

Figure 8.2 Unit for transferring data from a peripheral to a microprocessor

which distinguishes between an I/O and memory reference is generated and combined with either read or write control signals and the port address. This combination selects the required port and initiates the transfer of data.

The alternative method of approach to I/O operations assigns memory addresses to I/O ports and disposes of the requirement for two sets of read/write signals, one for I/O reference and the other for memory reference. There is, in this mode of operation, no distinction between I/O and memory addresses. It is not now necessary to provide specific I/O instructions, since the instructions provided by the instruction set for memory reference can now also be used for I/O operations. However, a portion of the memory space has to be allocated for I/O addresses.

The advantages of memory mapped I/O operations are that no new instructions are required for I/O operations. An I/O operation is indistinguishable from a normal memory access operation, and no additional I/O status pins are required on the processor chip. However, memory mapped I/O reduces the available memory space that can be used for memory reference, and, since memory is normally allocated in blocks, a single peripheral can occupy as much as 4 K bytes of memory space. Nevertheless, there is now greater software flexibility for the programmer since there are many more memory instructions available which can be used for I/O operations.

Some microprocessors, such as the 6800, are forced to use memory mapped I/O operations since there are no input and output instructions in the instruction set and the isolated I/O technique is not available. However, in the 8085A both modes of I/O operation are available.

In the 8085A, eight bits are used for I/O addressing. The machine generates the control signal $IO/\bar{M} = 1$ after the input instruction has been decoded, and this is combined with the control signal $\overline{RD}$ to produce the input read signal $\overline{IO/R}$, as illustrated in Figure 8.3(a). The address space is in the range 00 to FFH, and hence there are 256 locations available for addressing. Similarly, when writing into an output port, the address of the port is defined by eight address bits and lies in the range 00 to FFH. For an output write operation, the control signal $IO/\bar{M} = 1$ and is combined with the control signal $\overline{WR}$ to form the output write signal $\overline{IO/W}$, as shown in Figure 8.3(b). In all, it is possible for the processor to service 256 input ports and 256 output ports when employing the isolated I/O technique.

When employing isolated I/O, the full memory address space in the range 0000H–FFFFH is available for memory references.

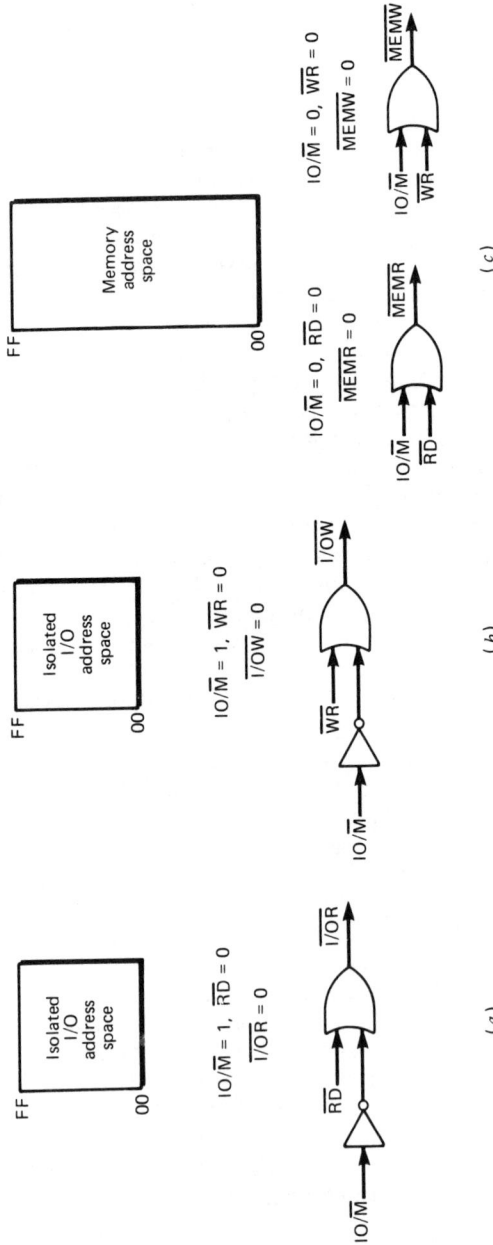

Figure 8.3 I/O and memory address space and the generation of I/O and memory control signals for isolated I/O techniques (a) I/O read (b) I/O write (c) Memory read or write

When addressing memory, the control signal $IO/\overline{M} = 0$ and this signal, in conjunction with either $\overline{RD}$ or $\overline{WR}$, is used to generate the memory read ($\overline{MEMR}$) and memory write ($\overline{MEMW}$) signals, as shown in Figure 8.3(c).

8.4 The input/output instructions for the 8085A

The two instructions provided in the 8085A instruction set for the isolated I/O mode have the mnemonic forms IN and OUT, and both of them are two-byte instructions. The first byte in each case is the instruction opcode which, when decoded, initiates an I/O operation, and the second byte is the address of the port which has to be read in the case of a data transfer into the processor and written into for an output transfer of data. For the IN instruction, when executed, the data is transferred from input port to accumulator while for the OUT instruction, data that has been placed in the accumulator is transferred to an output port.

The register transfer language statements and the hexadecimal representation for these two instructions are

IN $(DB)_{16}$ $(A) \leftarrow (data)$
OUT $(D3)_{16}$ $(data) \leftarrow (A)$

where (data) in the RTL statements refers to the data placed on the data bus in the first case by the specified input port, and in the second case data in the accumulator destined for a specified output port via the data bus.

8.5 The timing diagram for an isolated I/O operation in the 8085A

The timing diagram for an I/O read machine cycle is given in Figure 8.4 and is identical to that of an opcode fetch machine cycle except that the status signals generated during T_1 are $IO/\overline{M} = 1$, $S_1 = 1$ and $S_0 = 0$, and they identify the machine cycle as one in which data is transferred from an input port to the processor. A second difference is that a read cycle or, for that matter, a write cycle, always lasts for three clock periods or T-states rather than four to six T-states for an opcode fetch. At the end of T_3 the processor enters the next machine cycle.

The IN instruction requires three machine cycles for execution. The first is the opcode fetch, the second is a memory read, during which the port address is transferred from memory to the processor and is duplicated in the W and Z registers, while the third machine cycle, shown in Figure 8.4, is associated with the data transfer. During this last machine cycle, the 8-bit port address appears in duplicate on the upper and lower bytes of the address bus. A similar description can be given of the execution of the OUT instruction.

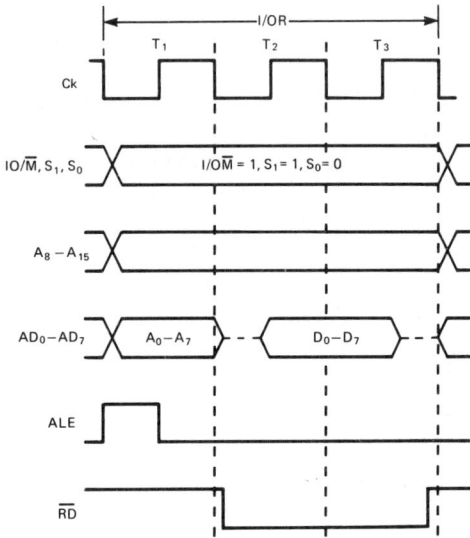

Figure 8.4 Timing diagram for an I/OR machine cycle in the 8085A

Timing diagrams for the I/O read and I/O write cycles are identical, except that a read control signal $\overline{RD}$ is generated by the processor when reading from an input port, and a write control signal $\overline{WR}$ is generated when the processor is writing into an output port. Additionally, the status signals S_1 and S_0 which are $S_1 S_0 = 10$ for I/O read become $S_1 S_0 = 01$ for an I/O write.

8.6 Design of peripheral selection logic

The methods used for the selection of a peripheral device are identical to those described in Section 3.13 of the chapter on Memory, where the selection of memory chips using memory address decoding techniques was described. Design of peripheral selection logic is governed by the number of peripherals and I/O ports associated with the microcomputer system. For example, if the system only requires a single input port and a single output port, then no address decoding is required. When the IN instruction is executed then $\overline{RD} = 0$ and $IO/\overline{M} = 1$ and the logical combination of these two signals can be used to generate the device strobe for releasing data from the input port, as shown in Figure 8.5(a). In exactly the same way it is also possible to generate the output device strobe when the OUT instruction is executed (see Figure 8.5(b)).

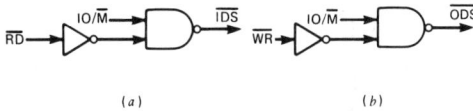

(a)   (b)

Figure 8.5 Selection of a single (a) input (b) output port

For more than one input or output port, address decoding is required, and the simplest technique available is the method of *linear selection*. An I/O address in the 8085A consists of eight bits A_0–A_7, or, alternatively, A_8–A_{15}, and a peripheral device is selected when only one of these address bits is held at 1. The device select strobe is now generated by a combination of $\overline{RD}$ or $\overline{WR}$ with $IO/\overline{M}$ and a single address bit, as indicated in Figure 8.6(a) and (b).

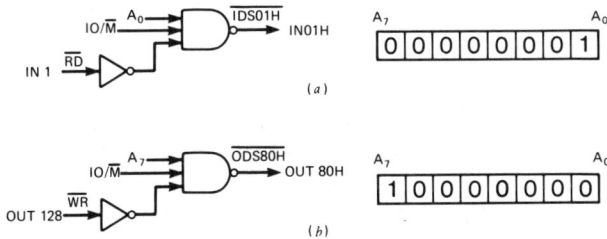

(a)

(b)

Figure 8.6 Linear selection of (a) an input port and (b) an output port

However, linear selection must be used with caution. If, for example, a simple programming error is made, resulting in the substitution of IN18H for IN08H, the execution of this instruction would result in the two address bits A_4 and A_3 simultaneously being equal to 1 and the consequence of the error would be the addressing of two input ports simultaneously. As far as the hardware is concerned, this addressing error would result in two sets of tri-state buffers attempting to drive the data bus simultaneously, and this could possibly cause permanent damage to the buffers.

Because there are only eight unique positions for a single one in the port address, the maximum number of addressable ports is 16, eight of which are input ports and the other eight output ports. Since decoding techniques are not required when this method is used, there is a considerable hardware saving which may be economically significant in a small microcomputing system.

It is common practice to use an address decoder for selecting the I/O ports that have to be accessed when either an IN or an OUT instruction appears in the program sequence. The scheme shown in Figure 8.7 provides access to eight input ports and eight output ports with the aid of two 3-to-8 line decoders. In addition, combinational logic is required to generate the $\overline{\text{I/OR}}$ and $\overline{\text{I/OW}}$ signals.

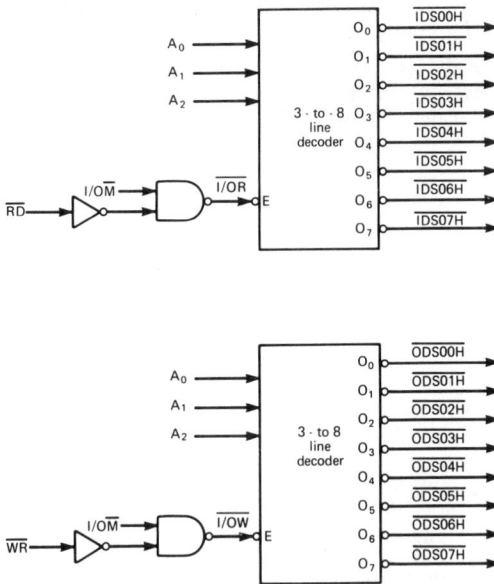

Figure 8.7 Non-absolute address decoding for eight input ports and eight output ports

Here is an example of non-absolute decoding, and there are a number of IN instructions that will generate the device strobe $\overline{\text{IDS00H}}$, because the address bits A_3–A_7 are not uniquely defined and can take a value of 0 or 1 when $A_0 = A_1 = A_2 = 0$. For example, any input instruction XXXXX000H will generate input device strobe $\overline{\text{IDS00H}}$ which, in turn, will release data on to the data bus.

Many devices now have multiple enable pins, and if these are available on the decoder it can eliminate the necessity for the combinational logic used in the decoding system shown in Figure 8.7. This point is illustrated in Figure 8.8, where the 3-to-8 line decoder is provided with two active low enables and one active high. The $\overline{\text{RD}}$ signal is connected to one of the active low enables while the other is grounded. The remaining enable pin, which is active high, is supplied by the IO/$\overline{\text{M}}$ signal.

To provide absolute address decoding for 16 input and 16 output ports, the system shown in Figure 8.9 can be used. The output of the two 4-to-16 line decoders will produce an active low output

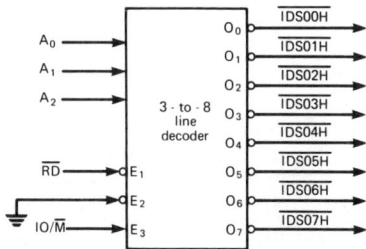

Figure 8.8 Non-absolute address decoding with combinational logic eliminated by multiple enables

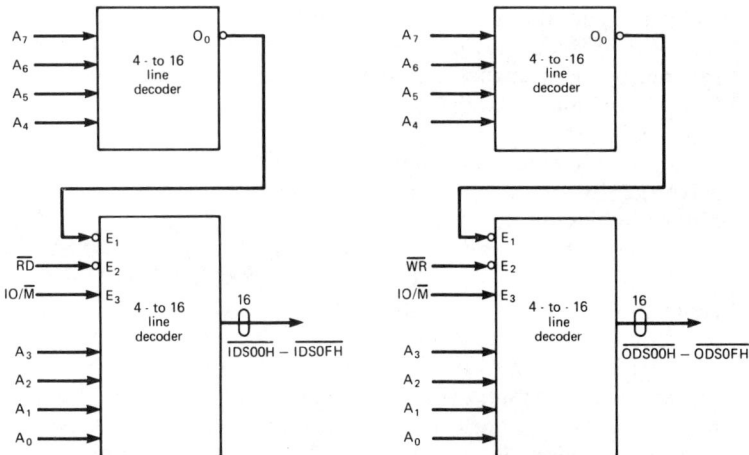

Figure 8.9 Absolute address decoding for 16 input and 16 output ports

signal at the terminal marked O_o when A_4–A_7 inclusive are all zero. This signal is used as one of the mutliple enables provided on the lower 4-to-16 line decoders, in conjunction with the $\overline{RD}$ or $\overline{WR}$ and IO/$\overline{M}$ signals.

Since there are eight address bits available for addressing I/O ports, it is possible with an 8085A processor to input data from 256 input ports and to output data to 256 output ports, a total of 512 ports in all. However, this would require a considerable amount of external hardware. For example, to process data from 256 input ports, a total of seventeen 4-to-16 line decoders are required, and to write to 256 output ports, a further seventeen 4-to-16 line decoders would be needed.

8.7 Memory mapped I/O operations

In the memory mapping technique, each I/O port represents a memory location in the memory map, and this implies that an I/O port will in these circumstances have a 16-bit address. The method allows transfers to take place from any one of the general purpose registers B, C, D, E, H and L, and the accumulator, to a peripheral uniquely defined by a memory address, via the data bus. Additionally, it is possible to transfer data from an input port via the data bus, and add it to the contents of the accumulator, leaving the result in the accumulator. For this type of operation, reading of data from an input port and writing data into an output port is controlled by the $\overline{MEMR}$ and $\overline{MEMW}$ signals respectively, which are generated from a combination of IO/$\overline{M}$ and $\overline{RD}$ or IO/$\overline{M}$ and $\overline{WR}$. Effectively, the memory mapping technique persuades the microprocessor that it is communicating with memory when, in fact, it is communicating with an I/O port uniquely defined by a 16-bit address.

A common method of memory mapping is illustrated in Figure 8.10(a). The memory map is split into two equal parts by the two binary values of the address bit A_{15}. The address range of the lower half of the address space 0000H–7FFFH is allocated to memory by making $A_{15} = 0$, while the upper half 8000H–FFFFH is allocated to I/O operations by making $A_{15} = 1$. Each section of the memory map is associated with 32K locations, and clearly it is a wasteful but nevertheless convenient allocation because a microcomputer system will never require 32K I/O ports. Since the microprocessor fetches the first instruction to be executed from memory location 0000H after it has been reset, it is convenient to set $A_{15} = 0$ when defining the memory space.

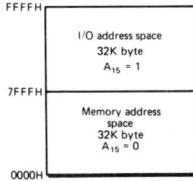

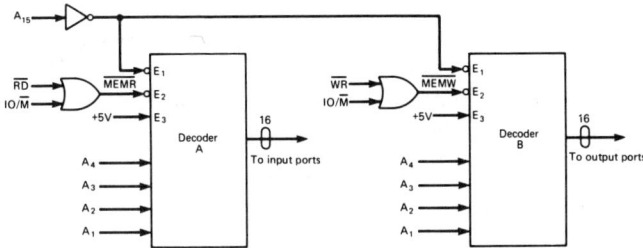

(b)

Figure 8.10 (a) The memory map for memory mapped I/O operations (b) Non-absolute memory mapped I/O device address decoding system

8.8 Port selection in a memory mapped system

Port and peripheral selection is not significantly different when memory mapped I/O is employed. The system shown in Figure 8.10(b) is typical of the techniques used. Two decoders are employed, decoder A being used to select 1-out-of-16 input ports, and decoder B to select 1-out-of-16 output ports. When the most significant address bit $A_{15} = 1$, peripherals are being addressed, and when $A_{15} = 0$, memory locations are either being read from or written into.

A_{15} is inverted and is used as an active low enable signal for both decoders. A second active low enable $\overline{MEMR}$ is generated from the logical combination of $\overline{RD}$ and $IO/\overline{M}$, which are both logical 0 for an input operation in a memory mapped I/O system. The 16 combinations of $A_3\ A_2\ A_1\ A_0$ are then used to select 1-out-of-16 input ports after decoding. The address bits A_4–A_{14} can take on any value during the peripheral selection process, consequently there are 2048 possible address combinations for selecting each of the 16 input ports, and none of these ports has a unique address.

The arrangements for selecting an output port are identical to those described above except that the second active low enable signal $\overline{\text{MEMW}}$ is generated from the logical combination of $\overline{\text{WR}}$ and IO/$\overline{\text{M}}$. When both of these signals are logical zero, $\overline{\text{MEMW}} = 0$ and decoder B is enabled, allowing any one of the 16 possible output ports to be selected, and as in the case of the input ports, none of them will have a unique address.

The system shown in Figure 8.10(b) can be turned into an absolute system, as illustrated in Figure 8.11. Input ports can now only be selected when $A_{15} = 1$, the address bits A_4–A_{14} are all logical zero, and $\overline{\text{MEMR}} = 0$. Each input port now has a unique address. Similarly, one of the 16 output ports is selected when $A_{15} = 1$, address bits A_4–A_{14} are all logical zero, and $\overline{\text{MEMW}} = 0$, thus allocating to each of these ports a unique address.

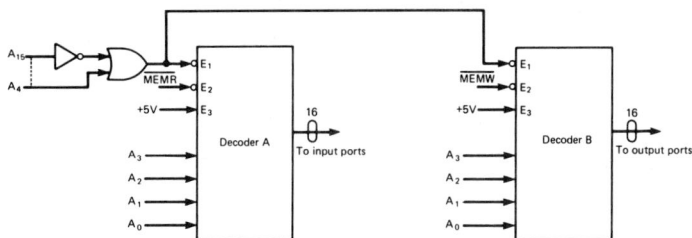

Figure 8.11 Absolute address decoding for memory mapped I/O devices

8.9 Assembly language instructions for memory mapped I/O devices

When memory mapped I/O is employed, instructions which reference memory can be used for transferring data between a peripheral and the microprocessor, and vice versa. This has been made possible by assigning a 16-bit address within the memory map to an I/O port. The register that constitutes the I/O port is now seen by the processor as a memory location.

The use of the memory mapped I/O technique results in increased programming flexibility and provides more than 40 instructions for I/O operations. By comparison, there are only two instructions available for I/O operations when the isolated I/O mode is used, namely IN and OUT.

A typical example of an instruction which can be used for implementing a data input from a memory mapped port is

MOV r, M   (r)←((H)(L))

The M indicates a transfer from a memory address allocated to a port, the address of the memory location being contained in the H and L registers. In this case the transfer does not involve the accumulator, data being transferred directly via the data bus to register r.

A second instruction that can be used in this way is

SUB M $(A) \leftarrow (A) - ((H)(L))$

The register transfer language statement defines the subtraction of the contents of the register whose address is specified by the contents of the H and L registers from the contents of the accumulator, the difference being returned to the accumulator after the subtraction has been performed in the ALU.

For memory mapped output operations, the following instruction may be used:

MOV M, r $((H)(L)) \leftarrow (r)$

When this instruction is executed the contents of register r are moved to the output port whose address is specified by the contents of registers H and L. Alternatively, the following move immediate instruction can be used for outputting data:

MVI M, data $((H)(L)) \leftarrow (byte\ 2)$

and the implementation of this instruction causes the data contained in byte 2 of the instruction to be written into the output port whose address is specified by the contents of the H and L registers.

8.10 Transfer of parallel data under program control

When data is to be transferred in parallel form from a peripheral to an 8-bit microprocessor, the data byte is placed in an 8-bit register called an input port. If the IN instruction is now executed, the eight bits in the port are simultaneously transferred to the accumulator in the processor. In the reverse direction, the data byte is first placed in the accumulator and is then transferred to the 8-bit output port or register when the OUT instruction is executed. The data transfers just described are carried out under program control in the isolated I/O mode.

Such a transfer can be either *unconditional* or *conditional*. For example, if data is to be transferred from an input port to the processor, the processor can assume that the peripheral has made fresh data available in the port after a specified time interval since the last data byte was transferred. The reader should observe that in this type of transfer the peripheral does not raise a flag to indicate to the processor that fresh data is available.

Alternatively, when the transfer is conditional, data can only be transferred after the peripheral has indicated the availability of fresh data by raising a DAV (data available) flag. The processor is now required to examine the DAV flag at specified time intervals to determine whether it has been raised. Clearly, processing time is lost during those time intervals when the flag is being examined.

A block diagram for this type of conditional transfer is given in Figure 8.12(a). The peripheral generates a strobe (STB) which

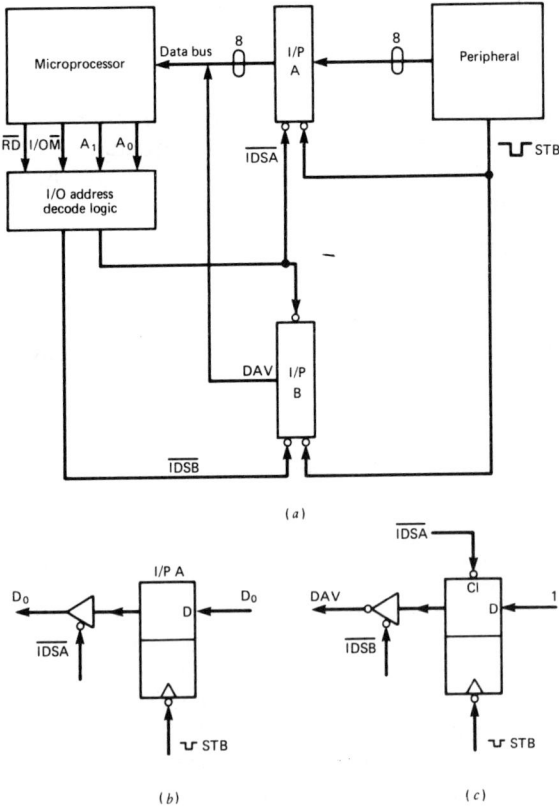

Figure 8.12 (a) Block diagram showing hardware requirements for a conditional transfer of data (b) Details of connections to I/P A (c) Details of connections for generation of $\overline{\text{DAV}}$ signal

serves two functions, first to latch the data into input port A (Figure 8.12(b)), and second, to latch the DAV signal into input port B. Examination of DAV is achieved by selecting input port B with the $\overline{\text{IDSB}}$ signal generated by the I/O address decode logic. If DAV = 1, the processor initiates the generation of $\overline{\text{IDSA}}$ which releases data from port A onto the data bus, and thus to the accumulator, and simultaneously resets DAV = 0.

The disadvantage of the conditional system just described is that the receipt of data by the processor is not acknowledged, and consequently the peripheral is not aware of the receipt of the data. This has led to the widely used *handshaking* transaction between processor and peripheral, which involves an acknowledgement of the receipt of the data and simultaneously defines the time at which the transfer is complete. Conceptually, the handshake mode is analogous to the despatch of an invitation with the letters RSVP attached. The person sending the invitation does not know that it has been received until the acknowledgement in response to the letter's RSVP is returned.

A state diagram describing a transfer in the handshake mode is given in Figure 8.13(a). When in the quiescent state (QS) the peripheral indicates that data is available by sending the DAV signal to the processor, thus initiating a transition to state S_1. While in this state the peripheral waits for the processor to acknowledge the DAV signal, which it does by returning the data acknowledge (DAA) signal to the peripheral. The DAA signal initiates a transfer to S_2, and in this state the data transfer takes place and DAV is set low by the peripheral. On completion of the transfer the processor sets DAA = 0, and a return to S_0, the quiescent state, occurs.

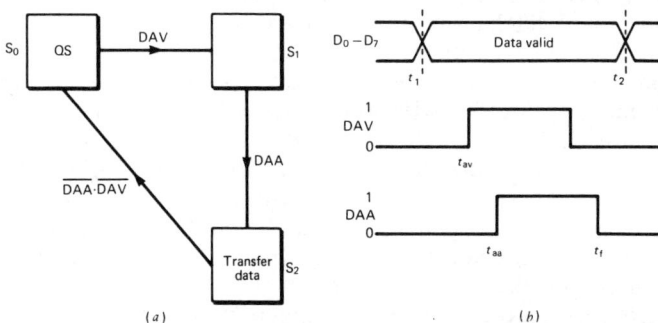

Figure 8.13 (a) State diagram for the handshake mode of operation (b) Timing diagram for the handshake

A timing diagram for the data transfer is shown in Figure 8.13(b). During the time period t_1 to t_2, the data output from the peripheral is stable. At time t_{av} the peripheral raises the DAV signal, and at time t_{aa} the processor raises the DAA signal. The transfer takes place in the time interval t_{aa} to t_f and is completed at time t_f when DAA goes low.

If a fault occurs during a handshake transfer, then an examination of the state diagram reveals the possibility of a 'lock-in'. For example, if DAA is not raised, then the handshake interface is permanently locked in state S_1. It is therefore advisable to have a *timeout* mechanism, capable of raising an alarm if the data transfer has not been completed within the time prescribed, this time being defined by a counter. It can be arranged that the counter, usually a processor register, will count clock pulses, and if the DAA signal has not been returned to the processor when the register count is zero, the alarm is raised.

The handshake mode of operation can be used to implement an asynchronous bidirectional transfer of data between peripheral and processor. The transition between states (Figure 8.13(a)) is controlled by the occurrence of an event such as the raising of the DAV signal, and is not in any way synchronized to the processor clock. This allows the development of an asynchronous interface between processor and peripheral, and because of the use of the handshake mode it accommodates devices that run at quite different speeds.

8.11 Polling

If a processor has to service more than one peripheral, then a subroutine has to be developed for testing the flag associated with each peripheral. Such a subroutine is termed a *polling* subroutine, and it has to be called at regular time intervals during the main program sequence. The time intervals are governed by the programmer's prior knowledge of the I/O requirements of individual peripherals.

Once the polling subroutine has been entered, the flags of the individual peripherals are examined in turn, as illustrated in Figure 8.14(a). The order of polling defines the order of peripheral priority. In this example, the flag of peripheral 4 is examined first and is therefore deemed to have the highest priority. Once the peripheral that is either available or is requesting service has been identified, its service routine is executed. At the end of the service routine the two actions shown in Figure 8.14(a) are possible.

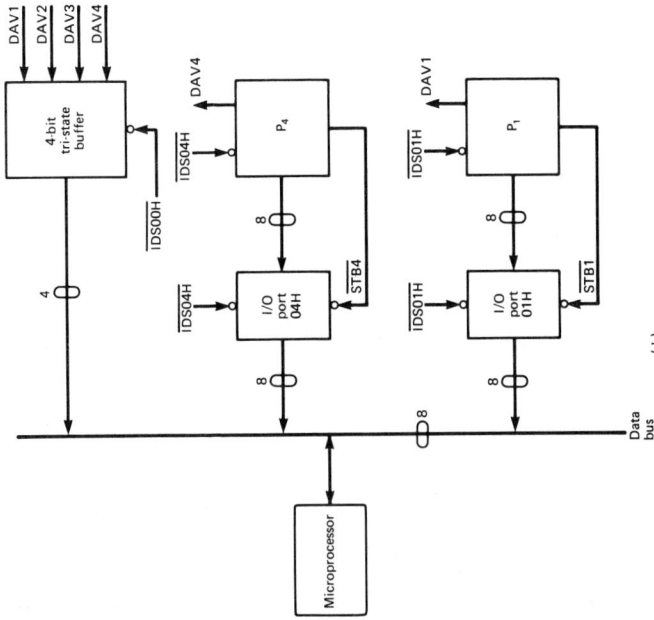

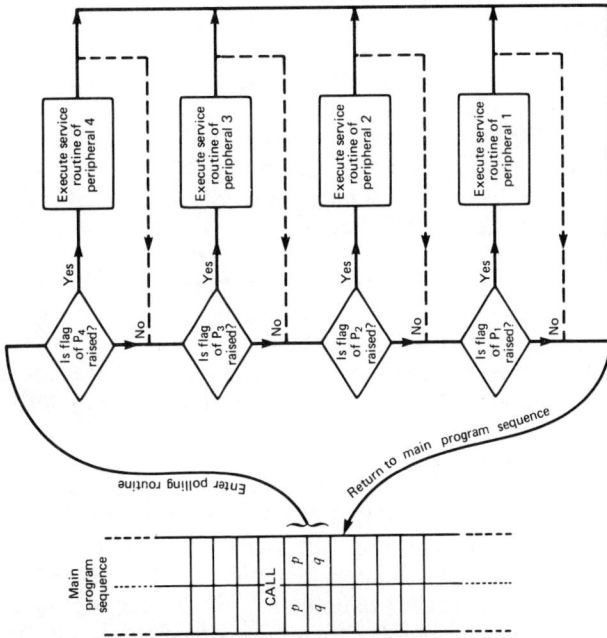

Figure 8.14 (a) Flowchart for polling sequence (b) Block diagram for transferring data between four peripherals and a microprocessor, using polling technique (P$_2$ and P$_3$ are similar to P$_1$ and P$_4$)

Control may be returned to the main program sequence without any further flags being examined. This implies that those peripherals at the head of the polling sequence will be examined more frequently than the ones at the bottom of the sequence, particularly if priority has been determined on the basis of the number of times a peripheral is likely to request service in a given time interval.

Alternatively, after the execution of the service routine, the polling sequence is continued, and all flags in the sequence are examined on each occasion the polling subroutine is called. Control is returned to the main program sequence only after all the peripherals have been polled.

If the first technique is employed, the lowest priority peripheral may be denied service for long time periods, but in the second method all peripherals, irrespective of priority, are polled at regular time intervals.

The disadvantage of the polling technique is that the flags of all peripherals have to be examined to determine their I/O availability, irrespective of whether they are requesting service or not. Every time a flag is examined, it requires a section of program which forms part of the polling subroutine, and of course requires time to execute. This is time wasted when the peripheral is not requesting service. The I/O overhead incurred in this manner becomes increasingly significant as the number of peripherals included in the polling sequence increases; in effect it represents processing time wasted.

An excessive I/O overhead may be eliminated by allowing data transfers to be initiated by a peripheral rather than the processor. This is the basis of the interrupt technique frequently employed in microcomputing systems and described in detail in the next chapter.

A considerable amount of hardware is required external to the processor and peripherals to implement a polling system. Each peripheral has to generate a strobe which serves the two functions described in Figures 8.12(b) and (c); namely, it generates the DAV flag and it latches the data provided by the peripheral into a tri-state register.

A microcomputing system for transferring data from four peripherals, P_1-P_4, to a microprocessor is shown in Figure 8.14(b). It is assumed in this diagram that the flag generating circuit is incorporated in the peripheral block. Initially, the 4-bit tri-state buffer is addressed via the I/O decode logic and the status of the four flags is transferred to the accumulator. The polling subroutine is now executed and each flag is examined in turn. If the flag

DAV_2 for P_2 is 1, then the corresponding service routine for that peripheral is called and executed. A polling subroutine for the system is shown below, in which it is arranged that all flags are examined once the subroutine is entered. A return to the main program sequence occurs after the peripheral with the lowest priority has been polled and serviced if necessary.

```
POLL:  IN 00H    ; Input flags
       CM SR 4   ; Transfer data if flag 4 is raised
       RAL       ; Rotate left
       CM SR 3   ; Transfer data if flag 3 is raised
       RAL       ; Rotate left
       CM SR 2   ; Transfer data if flag 2 is raised
       RAL       ; Rotate left
       CM SR 1   ; Transfer data if flag 1 is raised
       RET       ; Return
```

8.12 Peripheral interface chips

It is common practice among microprocessor manufacturers to provide support chips for their products which perform a variety of functions such as interfacing, control of direct memory access, and program interrupt control. All of these LSI chips are programmable, and the one of immediate interest from the I/O point of view is the 8255A Programmable Peripheral Interface, a highly flexible device which can be programmed to operate in three distinct modes and performs the interface function between a peripheral and the 8085A central processor.

The 8255A is a 40-pin chip with connections to and from the chip, as shown in Figure 8.15. Internally there are three ports, A, B and C, which can be used in either the input or output mode,

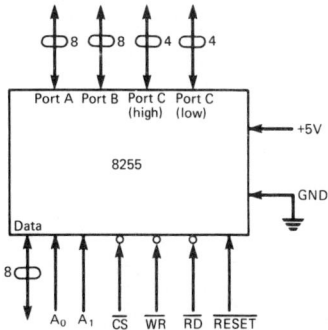

Figure 8.15 Pin connection to the programmable peripheral interface

and are able to carry both data and control information. Additionally, there is a data buffer which is connected to the system bus and carries data from the PPI to the processor, and vice versa.

A chip select signal $\overline{CS}$ allows the PPI to be selected by the processor and $\overline{WR}$ and $\overline{RD}$ inputs when active enable the processor to write into the PPI or, alternatively, allow it to read from the PPI. The address lines A_0 and A_1 are internally decoded on the chip so that 1-out-of-3 ports is selected during a read or write operation. When the RESET line is held at 1 the three ports A, B and C are operating as input ports and it is normal practice to wire this input to the RESET OUT pin of the 8085A, for obvious reasons.

The architecture of the 8255A is shown in Figure 8.16 and a truth table describing the basic operation of the PPI is given in Figure 8.17. Examination of the truth table reveals that when

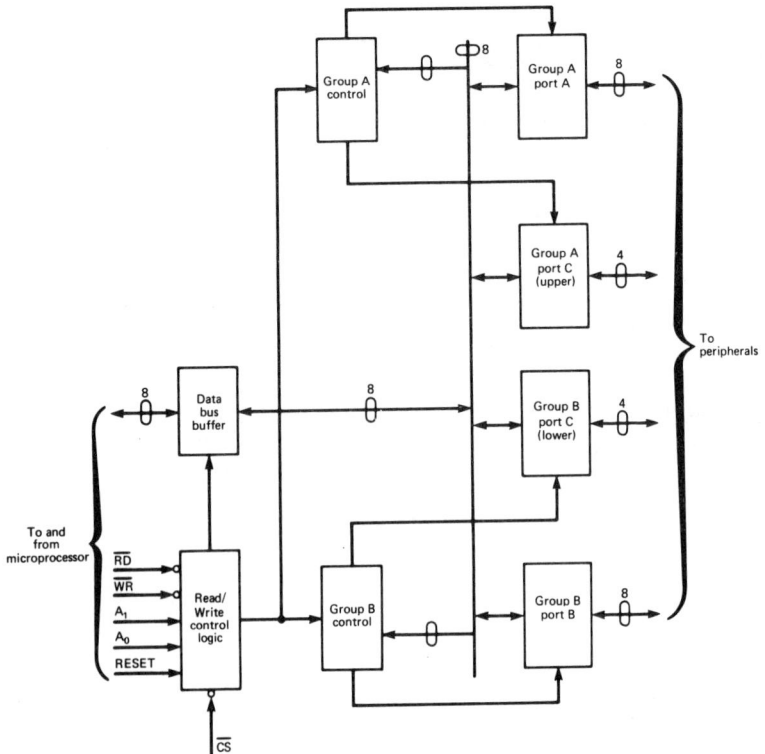

Figure 8.16 Architecture of the 8255A

A₁	A₀	R̄D̄	W̄R̄	C̄S̄	Action
0	0	0	1	0	Port A → Data bus (Read)
0	1	0	1	0	Port B → Data bus (Read)
1	0	0	1	0	Port C → Data bus (Read)
1	1	0	1	0	Forbidden combination
0	0	1	0	0	Data bus → Port A (Write)
0	1	1	0	0	Data bus → Port B (Write)
1	0	1	0	0	Data bus → Port C (Write)
1	1	1	0	0	Write mode control word to control register
X	X	X	X	1	8255A not selected
X	X	1	1	0	Data bus tri–stated

Figure 8.17 Basic truth table for the 8255A

$\overline{CS} = 0$, $\overline{WR} = 0$ and $\overline{RD} = 1$, the data placed on the system data bus by the processor is either sent to port A, B or C, or, alternatively, a control byte is directed to the control register which is incorporated with the read/write logic on the chip. The control byte serves two functions: (a) to output a single data bit through port C, or (b) to select the PPI mode of operation.

When the most significant bit of the control byte is 0, then the least significant four bits in the control register are used for defining a single-bit output on port C, as illustrated in the control register format shown in Figure 8.18. Three of these bits are used as an address to define which of the eight bits of port C is to be set or reset. The bit allocation shown in Figure 8.18 causes bit 2 of port C to take up the value of the least significant bit which, in this case, is 1.

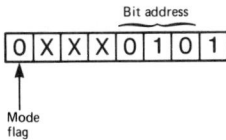

Bit address

0	X	X	X	0	1	0	1

↑
Mode
flag

Figure 8.18 Control register format for single-bit output

If the most significant bit in the control register is set to 1, the remaining seven bits are used to select one of the three modes of operation specified below:

Mode 0 – each group (A and B) of 12 I/O pins may be programmed in sets of four to be either input or output.

Mode 1 – each group may be programmed to have eight lines for either input or output, and the remaining four lines in the group are used for handshaking or interrupt control signals.

Mode 2 – this is the bidirectional mode, which is only available for group A. The eight lines of port A are used bidirectionally for either input or output, and five lines (including one borrowed from group B) are used for handshaking.

The format of the control word when used for mode selection is shown in Figure 8.19. For a control word 10011000 the PPI is operating in Mode 0, and Port A and the upper 4 bits of Port C are acting as input ports while Port B and the lower 4 bits of Port C are acting as output ports.

Mode Flag = 1	Select mode of Port A and upper 4 bits of Port C	Port A	Port C upper 4 bits	Select mode of Port B and lower 4 bits of Port C	Port B	Port C lower 4 bits
	Mode 0 = 00 Mode 1 = 01 Mode 2 = 1X	1 = Input 0 = Output		Mode 0 = 0 Mode 1 = 1	1 = Input 0 = Output	

Figure 8.19 Control register format for mode selection

When the control word is 1010X11X Port A is in the strobed output mode and Port B is in the strobed input mode. Now the PPI is operating in Mode 1. In this mode a port can be either input or output, but not both; nor is it necessary for both ports A and B to be simultaneously input or output ports. In Mode 1 six of the port C lines are used as control signals and provide the handshaking mode of operation while the remaining two can be connected as I/O lines. The connections to Port C are shown in Figure 8.20.

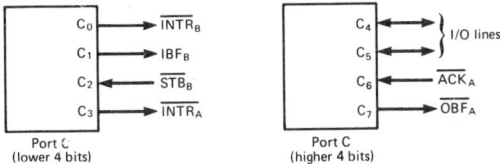

Figure 8.20 Connections to Port C in the handshaking mode

If a peripheral is connected to Port B when the PPI is operating in Mode 1, the strobe input signal $\overline{STB_B}$ is connected to C_2. On receipt of the strobe the port latches the data in the input buffer and returns the input buffer full (IBF_B) signal from C_1, thus acknowledging the receipt of the data. An interrupt request signal ($INTR_B$) is generated at C_0 and is used to indicate to the processor the presence of valid data in the buffer.

When data is sent to the PPI from the processor it is latched in an output buffer register. After latching, the output buffer full signal ($\overline{OBF_A}$) is generated and is transmitted to the peripheral; that is, in this case, connected to Port A, via pin C_7 on Port C. The peripheral accepts the data and generates an acknowledge signal ($\overline{ACK_A}$) which is returned to the PPI via pin C_6 on Port C. Pin C_3

on Port C carries an interrupt request signal which can, if necessary, be used to inform the processor that the PPI is ready to receive a new lot of data.

Mode 3 provides for a bidirectional transfer of data and is only available with Port A. It also supplies the handshaking signals that are required to maintain bus discipline. For a bidirectional transfer five of these signals are required. They are strobe in, input buffer full, output buffer full, acknowledge, and interrupt request. A control word format that places Port A in the bidirectional mode and Port B in the strobed input mode is 11XXX110. Mode 2 operation of Port A is identical to Mode 1 operation except that input and output transfers can be arranged to take place in any order.

8.13 Multi-function LSI devices

Besides the programmable peripheral interface, Intel also provide LSI multi-function devices such as the 8355, a 2 K byte ROM chip which additionally contains two general purpose I/O ports, each of which is eight bits wide, is bit programmable, and can be used in either the input or output mode. A second somewhat similar multi-function device is the 8155, an LSI circuit containing 256×8 bits of R/W memory, three programmable I/O ports and a 14-bit programmable counter. As an example of this type of multi-function device a description of the I/O facilities and operation of the 8355 follows, and its block diagram appears in Figure 8.21.

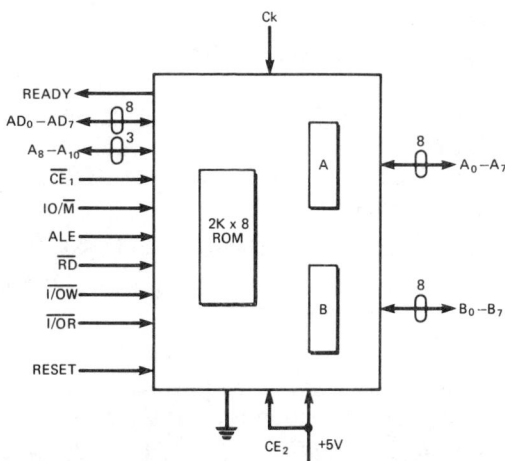

Figure 8.21 Block diagram for the 8355 ROM with I/O

The 8355 has two I/O ports A and B, and two data direction registers DDR A and DDR B, both of them eight bits wide. The DDRs determine the I/O status of each bit of the two ports. A 0 in the DDR specifies an input mode and a 1 specifies an output mode. If DDR A is programmed so that all bits in the register are 1s then Port A has been programmed as an output port, and if DDR B is programmed so that all bits in the register are 0s, then Port B has been programmed as an input port. The bits in the DDRs are individually programmable so that some port bits can operate in the input mode while others are in the output mode.

The I/O section of the chip is addressed by the latched values of AD_0–AD_7. Addresses are allocated to the DDRs to allow them to be programmed, and also to the ports so that data can either be written into or read from them when addressed.

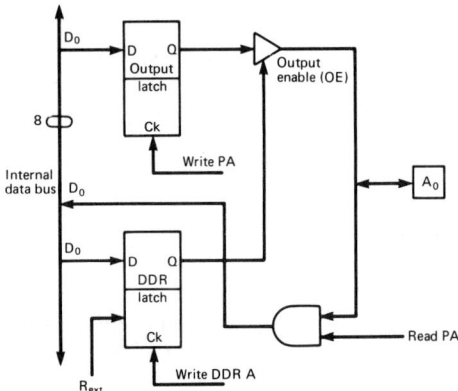

Figure 8.22 DDR and port arrangement for the 8355

To clarify the function of the I/O ports and DDRs the configuration of one bit of Port A and the corresponding bit of DDR A is shown in Figure 8.22. A 1 is clocked into the DDR latch by the write DDR A signal, which is given by

Write DDR A = $\overline{\text{IO/W}}$. CE_2 . CE_1 . (DDR A address)

and an output to A_0 now occurs when OE = 1. The output latch is operated by the write PA signal, which is given by

Write PA = $\overline{\text{IO/W}}$. CE_2 . $\overline{CE_1}$. (Port A address)

The microprocessor reads the contents of the port when the read PA signal is present. The logical equation for this signal is

Read PA = $\{[\text{IO/M} . \overline{RD} + \overline{\text{IO/R}}] . CE_2 . \overline{CE_{11}}$. (Port A address)$\}$

An example of a program listing is given below for programming Ports A and B as output ports with even digits in the ports displaying 1s and odd digits displaying 0s.

```
MVI A FFH    ;  Load DDR word into accumulator
OUT 02H      ;  Write into port A DDR
OUT 03H      ;  Write into port B DDR
MVI A AAH    ;  Load acc. with output word
OUT 00H      ;  Write to port A
OUT 01H      ;  Write to port B
HLT          ;  Halt
```

8.14 Serial transfer of data

Data transfer between processor and peripheral can be in either serial or parallel form. For an 8-bit processor using parallel transfer the transmission channel consists of eight parallel conductors. When using serial transfer techniques the data and control bits are organized into a group of bits referred to as a *character* and are transmitted sequentially down a single line. Parallel transfer is faster and requires considerable investment for anything but the shortest of transmission channels. Besides the cost of the cable, line drivers for converting T^2L levels into signals suitable for driving a transmission line are required, and also line receivers are needed at the receiving end for converting the received signal back to the required T^2L level. Hence parallel transfer is reserved for short distances where speed is essential. By comparison, serial transfer is considerably slower and cheaper. Serial transfer is also used when the peripheral is inherently serial in operation or, alternatively, when the computer is linked to the peripheral via the telephone system.

Figure 8.23 Serial interface between microprocessor and peripheral separated by a considerable distance

A block diagram for a serial interface between a microprocessor and a peripheral separated by a considerable distance is illustrated in Figure 8.23. Interface circuitry is required at both ends of the transmission channel and provides the following two functions:

(a) conversion of data format (serial/parallel or parallel/serial),
(b) electrical transformation of T^2L levels to those levels appropriate to the transmission channel, and vice versa.

There are three possible ways of transmitting data between a microprocessor and a peripheral:

(a) *Full duplex* (FDX) – data transmitted in two directions simultaneously. This requires two channels.
(b) *Half duplex* (HDX) – data transmitted in both directions but only one way at a time.
(c) *Simplex* – data transmitted one way only.

Methods (b) and (c) both require a single channel only.

Transmission of data in serial form can be either *asynchronous* or *synchronous*. In asynchronous systems data is not transmitted continuously but only when it is available. Such a system is event-driven, where the event which initiates transmission is the formation of a character at the transmitting end. When data is not available the line is quiescent or idle and is permanently held in the logical 1 condition. For a synchronous system the receiver and transmitter share a common clock frequency and the transmission rate is governed by the magnitude of this frequency. If there is a large distance between the transmitter and receiver, clock signals are generated independently at both ends of the transmission system and a clock synchronization signal is supplied periodically by the transmitter to ensure the synchronization of the two clocks.

Another important aspect of serial transmission is the transmission rate, measured in bits/sec. and frequently referred to as the *baud rate*, a term which finds wide usage in the field of communication engineering. Typical bit rates used in practice are 110, 150, 300, 600, 1200, 2400, 4800 and 9600. The slower transmission rates are normally associated with electromechanical devices while the fastest bit rates are generally used with devices that tend to be purely electronic in nature.

8.15 Asynchronous serial data transfer

Asynchronous transfer techniques are frequently used with electromechanical devices that operate at the lower end of the bit rate spectrum. A group of 11 bits constitutes a character, which is made up of a start bit, seven data bits, a parity bit and two stop bits. A typical character format is shown in Figure 8.24. The seven data bits are normally coded in ASCII, the American Standard Code for Information Interchange.

The ASCII code is shown tabulated in Figure 8.25. Notice that all codes in the range 00 to 1FH are control codes, those in the

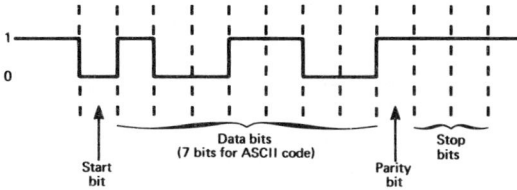

Figure 8.24 Character format for asynchronous serial data

Hex	ASCII	Hex	ASCII	Hex	ASCII	Hex	ASCII	Hex	ASCII	Hex	ASCII	Hex	ASCII	Hex	ASCII
00	NUL	10	DLE	20	SP	30	0	40	@	50	P	60	`	70	p
01	SOH	11	DC_1	21	!	31	1	41	A	51	Q	61	a	71	q
02	STX	12	DC_2	22	"	32	2	42	B	52	R	62	b	72	r
03	ETX	13	DC_3	23	£(#)	33	3	43	C	53	S	63	c	73	s
04	EOT	14	DC_4	24	$	34	4	44	D	54	T	64	d	74	t
05	ENQ	15	NAK	25	%	35	5	45	E	55	U	65	e	75	u
06	ACK	16	SYN	26	&	36	6	46	F	56	V	66	f	76	v
07	BEL	17	ETB	27	'	37	7	47	G	57	W	67	g	77	w
08	BS	18	CAN	28	(	38	8	48	H	58	X	68	h	78	x
09	HT	19	EM	29	)	39	9	49	I	59	Y	69	i	79	y
0A	LF	1A	SUB	2A	*	3A	:	4A	J	5A	Z	6A	j	7A	z
0B	VT	1B	ESC	2B	+	3B	;	4B	K	5B	[	6B	k	7B	{
0C	FF	1C	FS	2C	,	3C	<	4C	L	5C	\	6C	l	7C		
0D	CR	1D	GS	2D	−	3D	=	4D	M	5D	]	6D	m	7D	}
0E	SO	1E	RS	2E	.	3E	>	4E	N	5E	∧ (↑)	6E	n	7E	~
0F	SI	1F	US	2F	/	3F	?	4F	O	5F	− (←)	6F	o	7F	DEL

Figure 8.25 American Standard Code for Information Interchange (ASCII); seven bits with high order eighth bit (parity) set to zero

range 20 to 3FH are either special codes or represent numerical values, those in the range 40 to 5FH represent upper-case alphabet, and those in the range 60 to 7FH represent lower-case alphabet. Control codes are easily recognized by the microprocessor since the two most significant digits in these codes are always 0.

There are four basic characteristics that define a character and which enable its presence on the line to be detected by the interface circuitry:

(a) When there is no data available the line is held at logic 1. The advantage of this convention is that any break in the transmission line causes the logic 1 to disappear and hence indicates the presence of a fault.

(b) The start of a character is identified by the 1→0 transition of the start bit.
(c) The first and least significant data bit follows immediately after the start bit.
(d) After the parity bit there follow two stop bits, both having a value of logic 1.

The start bit and the two stop bits *frame* the 7-bit ASCII code and its associated parity bit.

Although the transfer of data is asynchronous, clock signals are required at the transmitting and receiving ends. At the transmitting end a PISO shift register has to be clocked to release the character to be transmitted onto the line, while at the receiving end a clock is required to clock the data from the line into a SIPO shift register. The receiver clock has to be generated from a knowledge of the nominal bit rate and by sensing the 0→1 and 1→0 data transitions on the line.

It is normal practice to sample the receiving end bit stream at a rate 16 times the nominal bit rate. Every time a change in data level is detected the receiver clock is reset to zero. Half-way through the nominal bit time, after eight cycles of the sampling frequency, the receiver clock is set to 1. When next a data transition occurs, or, in the absence of a data transition, after 16 cycles of the sampling frequency, the receiver clock is again reset to zero, thus generating one complete cycle of the receiver clock. The generated clock can now be used to clock data into the receiver shift register.

8.16 Synchronous serial data transfer

In a serial transfer system using synchronous techniques, each character consists of 11 bits, of which four carry no information at all. Three bits are used to frame the character and one bit is used for error detection. To improve transmission efficiency and to achieve faster data transmission, synchronous serial transmission is employed where long blocks of data containing many characters are transmitted. There are no character framing bits employed in this technique, so that each individual character contains a large information content. However, each block is preceded by a synchronizing sequence of bits which are used to identify the start of the block. Because the data block transmitted is long the overhead penalty incurred by comparison with the asynchronous technique is small.

A simplified transmission format is shown in Figure 8.26. In the case illustrated in this diagram the block is started with three identical SYN sequences. One of the functions of the receiver is to search for the synchronizing sequence by comparing the incoming bit sequence with the synchronizing pattern stored in memory. When coincidence of the two bit patterns has been established the start of the block has been identified.

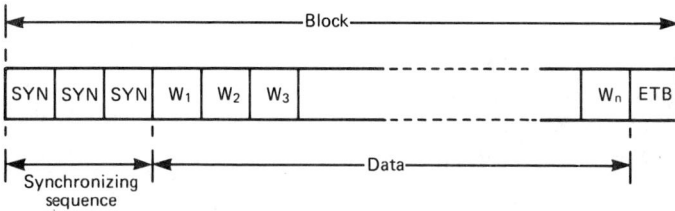

Figure 8.26 Synchronous serial transfer data format

Because the transmission channel is corrupted by noise, one or more bit errors may occur in the synchronizing sequence, SYN, and recognition of this sequence, and hence the start of the block, is missed. For this reason it is common practice to precede the start of the data with a number of synchronizing sequences to counter the occurrence of such a random error by ensuring that at least one SYN sequence remains uncorrupted.

If, for some reason, the receiving end falls out of synchronism with the transmitting end so that the receiver loses its correct position within a data block, incorrect words can be formed by starting midway through one word and entering the next word, as illustrated in Figure 8.27. To recover synchronism between transmitter and receiver, if such a fault should occur, the SYN character can be introduced at certain specified intervals within a block. When the SYN signal is recognized by the receiver, synchronism between transmitter and receiver will have been restored and data will be received again, starting from the correct bit.

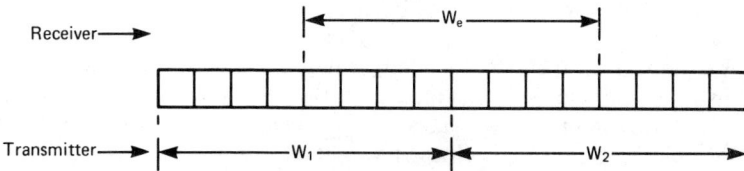

Figure 8.27 Formation of error word W_e due to lack of synchronization between transmitting and receiving ends

The ASCII code has a synchronizing character SYN and an examination of the tabulation of this code in Figure 8.25 reveals that SYN is represented by 16H. The corresponding binary format is 10010110, where the most significant bit is the parity bit, in this case 1, to provide even parity, and the remaining seven bits represent the control signal SYN.

8.17 Protocol

The set of rules that govern the transfer of data between two communicating devices in a synchronous serial system is called the *protocol*. It will be recalled that in the ASCII code those combinations in the hexadecimal range 00–1FH were specified as control codes and a limited number of these can be identified as those which control transmission of data. These codes, with their functions, are listed in Figure 8.28.

Code HD	Symbol	Function
00	NUL	All the 0's
01	SOH	Indicates start of header field
02	STX	Indicates start of text
03	ETX	Indicates end of text
04	EOT	Termination of transmission
05	ENQ	Enquire if terminal is on
06	ACK	Informs Tx of receipt of error free data
10	DLE	Data link escape
15	NACK	Informs Tx of receipt of data containing errors
16	SYN	Establishes bit and character synchronism
17	ETB	Indicates the end of block of data

Figure 8.28 ASCII characters for control of communication

There are two types of protocol (a) character oriented and (b) bit oriented. Character oriented protocols are based on the binary representation of a set of characters. The ASCII code, tabulated in Figure 8.25, is a set of 128 characters, each of which is represented by seven bits, and a number of these characters can be used as the protocol for a data communication system.

The function of SYN has been described previously. After the receiver has recognized the sequence of SYN characters provided by the transmitter, the line can go into the synchronized idle state in which data is not transmitted but bit and character synchronization are maintained. This, in effect, is the equivalent of transmitting the NUL character.

Alternatively, after the recognition of SYN, a message may be transmitted. Messages have a well-defined format. They consist of

three fields, header, text and error check. The header field
contains the address of the device with which the transmitter
wishes to communicate and also any other control bits that may be
required. This field is preceded by the start of header (S0H)
character and is terminated by the start of the text character
(STX). The STX character separates the header from the text, as
illustrated in Figure 8.29. Text then follows and can contain any
ASCII combination with the exception of those reserved for the
protocol. At the end of text, the end of text character (ETX) is
transmitted, and is followed by the block check character (BCC)
whose function is error detection.

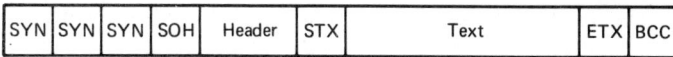

SYN	SYN	SYN	SOH	Header	STX	Text	ETX	BCC

Figure 8.29 Typical message format for a character oriented protocol

The receiver determines independently its own BCC count and
compares it with the transmitted BCC. If the two are in
agreement, the receiver returns the error-free acknowledge
character (ACK) to the transmitter. In the case of a disagreement,
the receiver responds to the transmission by sending a negative
acknowledgement (NACK) which indicates the presence of errors
to the transmitter and is an invitation to re-transmit. The
transmitter will then re-transmit until such time as ACK is
received, or until the operators have decided there is a fault in the
transmission system.

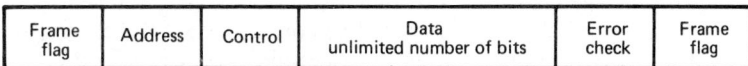

Frame flag	Address	Control	Data unlimited number of bits	Error check	Frame flag

Figure 8.30 Frame format for a bit oriented protocol

In data communication systems employing a bit oriented
protocol, characters are not used in the control field and they
allow the transmission of bit streams of any length. As in the
character oriented protocol, there is a specified message format,
called a frame, which contains four contiguous fields: address,
control, data and error check. The beginning and end of the frame
are defined by frame flags. A typical example of the frame format
is shown in Figure 8.30. Any further discussion of bit oriented
protocols is not within the scope of this volume and the reader is
advised to consult the many text books now available on data
communication systems for further information.

8.18 The universal asynchronous receiver/ transmitter (UART)

The device used when an asynchronous serial mode of data transfer is required between microprocessor and peripheral is the universal asynchronous receiver/transmitter. Such a device provides facilities for full duplex working and contains both transmitter and receiver sections. Both the microprocessor and the peripheral are connected in parallel with a UART, and one of these devices is required at each end of the communication link.

Assuming that the processor is sending data to a peripheral, then the transmitting section of the UART at the microprocessor end of the communication link receives the data in parallel form and converts it into serial form. Additionally, the transmitter provides start and stop bits and will, if required, generate a parity bit, thus forming a conventional 11-bit character which can now be transmitted along the line. At the receiving end the receiving section of the UART converts the serial data into parallel form after the removal of the start and stop bits and after parity has been checked. The 8-bit word is then transferred to the peripheral in parallel form.

Since a UART is designed for full duplex working it is equally possible to transfer data from the peripheral to the processor utilizing the transmitter section of the UART at the peripheral end of the communication link and the receiving section of the UART at the other end. A generalized block diagram showing the receiving and transmitting sections of a UART is given in Figure 8.31.

To initialize the transmitting section, the control register is programmed by the processor. The control byte entered into this register governs the mode of operation of the UART. For example, it will control the length of the character, whether even, odd or no parity is to be used, the baud rate, and the number of stop bits to be added to the data. The transmitter buffer register (a PIPO) accepts a data byte from the processor when the status flag indicates that the transmitter buffer is empty. This byte is then transferred to a shift register (a PISO) and simultaneously the parity generator provides a parity bit and adds it with the start and stop bits to the data in the shift register to form a character. This character is now ready for transmission and is clocked out of the shift register onto the communication link. A system of *double buffering* is employed, which allows fresh data to be loaded into the transmitter buffer when the status flag indicates that it is empty.

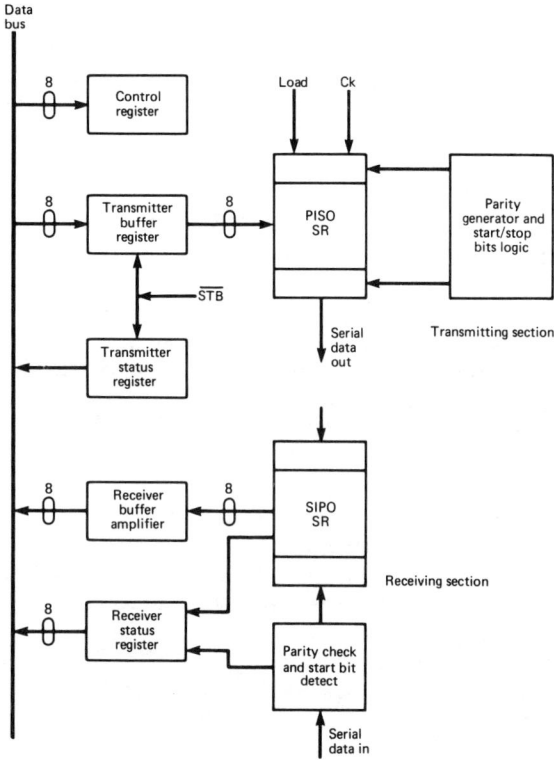

Figure 8.31 Block diagram of a conventional UART

At the receiving end, the receiver shift register receives the character in serial form after the start bit has been detected and the parity has been checked for the presence of errors. The detection of the start bit is used to synchronize the UART with the incoming bits for the time duration of the character, as previously described in Section 8.15. Additionally, the receiving section will provide status flags that indicate whether there is either a *framing* or *overrun* error. A framing error flag will be raised when the communication link does not return to the '1' condition at the end of a character, and an overrun flag is raised if a second character is received before the previous one has been read. When the error checks have been made, the data, if correct, is transferred to the receiver buffer register and thence to the processor in parallel form.

Facilities are frequently provided for conditional (handshake) or unconditional transfers of data. If the handshaking technique is employed, a data available flag (DAV) is raised. This can be examined by the processor and, if present, the processor can input the data. In some cases a programmable baud rate generator is included in the transmitter section of the UART and allows the selection of a number of bit rates.

8.19 Communication techniques with long transmission channels

For all but the shortest lengths of cable (less than 3.0 feet) standard TTL gates are incapable of driving the line that connects the microprocessor to the peripheral. Also, if the connection between these two devices exists in a noisy environment, corruption of the transmission channel by noise may occur. It is then possible that the noise margins available with the commonly used gates may not be sufficient and large induced noise spikes on the channel will lead to errors in the transmitted data. For this reason it is common practice to interpose a line driver and a receiver between processor and peripheral. The line driver converts the output TTL signals into levels suitable for transmission down a long line, while at the receiving end the receiver converts the received signals back to the appropriate TTL levels suitable for inputting to either processor or peripheral. A block diagram illustrating the transmission arrangements is shown in Figure 8.32.

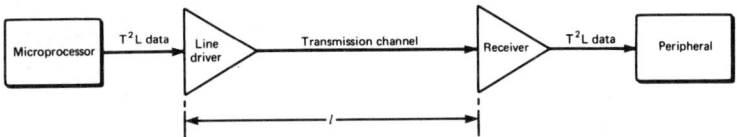

Figure 8.32 Block diagram showing transmission arrangements between a microprocessor and a peripheral

As the distance l between processor and peripheral increases, attention has to be paid to the method of interconnection. An important quantity associated with the line is the *maximum allowable bit rate*. As l increases the maximum allowable bit rate falls off in value. Factors which are important and influence this quantity are the R, L and C of the cable, the types of line driver and receiver employed, and the noise environment of the channel.

Cable capacitance C increases with the length of the line and constitutes an additional load on the driver, which results in a reduction in the amplitude of the received signal. The resistance R of the cable also increases with length, producing a voltage drop along the line and thus providing an additional limitation to the magnitude of the received signal.

The transmission can also be corrupted by electrostatically and electromagnetically induced noise signals which may significantly alter the received data and result in an increased bit error rate.

The effects described in the last two paragraphs can in the limit reduce the bit rate to zero. When this point has been reached *modems* have to be introduced at both ends of the channel. These devices accept digital waveforms and convert them into signals suitable for transmission over an analogue channel. They can also receive analogue signals from a distant modem and convert them back to their original form.

Transmission line effects may also occur on the channel. Reflections of the transmitted voltage can be generated at the receiving end and this may result in a significant reduction in the amplitude of the receiving end voltage. This problem can be avoided by terminating the line in a resistance value equal to the characteristic impedance of the line.

Finally, the receiving and transmitting ends of the transmission channel may have independent earth connections. This may lead to a potential difference between these connections due to the finite ground resistance between the two earth points. The potential difference may be substantial and acts as an offset voltage at the receiving end, which may lead to an incorrect logic interpretation of the received voltage.

There are three commonly used methods of interconnection:

(1) single-ended,
(2) unbalanced differential, and
(3) balanced differential.

For the single-ended case the driver is connected to the receiver by a single wire and there is no earth return wire. The transmission system is illustrated in Figure 8.33(*a*). The driver and the receiver are earthed independently at the transmitting and receiving ends. This has the advantage that only one wire is required per channel. However, the system performance may be less than satisfactory for two reasons:

(1) the presence of induced noise voltage v_n on the line, and
(2) the difference in potential between the two earth points A and B.

The voltage at the receiver v_r is given by the equation:

$$v_r = v_t \pm v_g \pm v_n$$

where v_t is the transmitting voltage, v_g is the difference in potential between points A and B, and where it is assumed that the line is loss free. The presence of the noise and ground voltages in this equation may result in a sufficiently large reduction in the amplitude of the received signal to lead to an incorrect interpretation of the binary signals on the line.

With the unbalanced differential line connection shown in Figure 8.33(b) there is a single earth return wire which may be common to a number of transmission channels. At the receiving end the receiver has a differential input that senses the potential difference between the two channel wires. Noise voltages induced in these two wires at any given instant will be approximately identical in magnitude and polarity and, because of the common mode rejection properties of the differentially connected receiver,

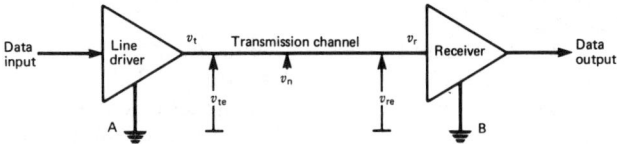

(a)

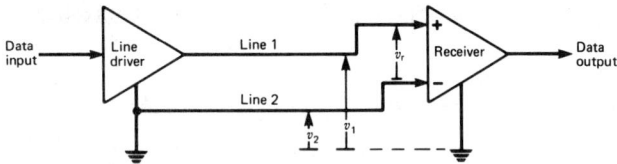

(b)

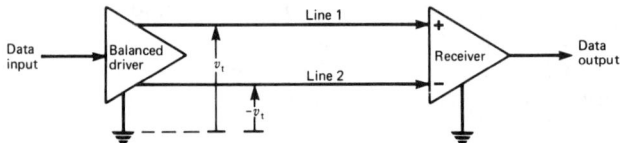

(c)

Figure 8.33 (a) Single-ended transmission system (b) Unbalanced differential connection (c) Balanced differential connection

the noise at the output of the receiver is virtually eliminated. Considering the effects of induced noise voltages only, the potential of line 1 with respect to ground is:

$$v_1 = v_t \pm v_n$$

and the potential of line 2 with respect to earth is

$$v_2 = \pm v_n$$

The input to the receiver is

$$v_r = v_1 - v_2 = v_t \pm v_n \pm v_n$$

and since at any instant the v_n's in the above equation will have the same sign

$$v_r = v_t$$

The noise voltages present at the receiver input cancel because of the differential connection and hence effectively eliminate noise at the output of the receiver. Similarly, it can be shown that any offset voltage due to different earth potentials at the two ends of the channel is eliminated by the differential connection at the receiver input.

The most effective interconnection uses a balanced driver, a receiver with a differential input and a two wire connection (Figure 8.33(c)). The driver has two outputs, $+v_t$ and $-v_t$, the voltage between the wires being $2v_t$, and thus a greater signal attenuation can be tolerated without a loss of data at the receiving end; consequently the length of the interconnection between driver and receiver can be increased. To reduce the effects of magnetic induction the two wires are twisted together and, because of common mode rejection at the receiver input, noise voltages and differences in earth potential are effectively eliminated at the receiver output. Also, if the cable pairs are fed from balanced drivers, crosstalk between adjacent channels is significantly reduced.

8.20 The serial I/O pins of the 8085A processor

The 8085A processor has special serial input and output pins named SID (serial input data) and SOD (serial output data). There are two instructions associated with these pins; first, RIM (read interrupt mask) associated with the SID pin, and second, SIM (set interrupt mask) associated with the SOD pin.

Input data appearing at the SID pin of the processor is transferred to bit 7 of the accumulator when the RIM instruction is executed. To output data from the SOD pin two requirements have to be met; first the data is placed in bit 7 of the accumulator, and a logic 1 is placed in bit 6 of the accumulator. When these requirements are satisfied the SIM instruction is executed and the logic 1 in bit 6 enables the SOD line, thus allowing the data in bit 7 of the accumulator to be transferred to the SOD pin.

Besides allowing the serial input and output of data, the SIM and RIM instructions are also associated with the machine interrupts. This aspect of these two instructions is dealt with in the chapter on interrupts.

It will be recalled that a character to be transmitted serially consists of a start bit (0), a 7-bit ASCII code and its associated parity bit, and two stop bits (1's); in all, 11 bits. A program listing is given below, which outputs a total of 15 characters in serial form and which is designed to generate the start bit and the two stop bits for each character. The 7-bit ASCII codes and their associated parity bits are stored in consecutive memory locations which are accessed by the pointer placed in the H and L registers. After the output of a character, a delay subroutine is called to ensure the correct time interval between a pair of consecutive bits.

```
SRDO:   LXI     H,4000H     ; Load starting address of data
                              in HL register
        MVI     D,0FH       ; Load no. of characters in D
                              reg.
LOOP:   MVI     A,40H       ; Load start bit in acc. and set
                              SOD enable
        SIM                 ; Output start bit
        MOV     B,M         ; Transfer data to B reg.
        MVI     C,0AH       ; Load character bit counter
CHCT:   MOV     A,B         ; Transfer data to acc.
        STC                 ; Set stop bit
        RAR                 ; Move bit to be transmitted in
                              to Carry.
        MOV     B,A         ; Transfer contents of acc. to B
                              reg.
        RAR                 ; Move bit to be transmitted
                              into b7
        ANI     80H         ; Mask all but b7 in acc.
        ORI     40H         ; Set SOD enable
        SIM                 ; Output data bit
```

```
DCR   C        ;  Decrement character bit
                  counter
JNZ   CHCT     ;
INR   H        ;  Increment pointer
DCR   D        ;  Decrement character count
JNZ   LOOP     ;
RET
```

Problems

8.1 Describe the difference between

(a) program controlled I/O,
(b) interrupt driven I/O, and
(c) DMA transfer.

Discuss the advantages of interrupt driven I/O in comparison with program controlled I/O.

8.2 Give a brief account of the techniques employed in linear and decoded selection of I/O devices, and discuss the relative merits of the two methods. Develop the logic circuitry required for a linear selection system which is designed to differentiate between five ports, two of them output ports and three of them input ports. Indicate the instructions that will be required to create the appropriate device select pulses.

8.3 Using 3-to-8 line decoders having two active low enables and one active high, develop a decoding system which will generate 12 input device select pulses and eight output device select pulses making use of the appropriate 8085A control signals. Give the address of each port.

8.4 Data is to be transferred from five peripherals to an 8085A processor. The data ready flags are provided by the five most significant bits of a status word. Develop a software routine for polling each of the peripherals and, additionally, write a subroutine for transferring a block of data of defined length from one of the peripherals to a specified sequence of locations in memory.

8.5 Readings of an analogue quantity in digital form are to be processed by an 8085A processor. Readings are taken every 100 ms and at the end of 1 minute the readings are averaged and placed in memory location 1000H. The processor initiates conversion with an SOC (start of conversion) *pulse*. Develop a suitable software routine. It can be assumed that the conversion time is small in comparison with the time between readings.

8.6   Write a program that will give a scale-of-10 count on a seven-segment display. Assume that the 8085A is operating at an internal clock frequency of 3.125 MHz and provide an appropriate timing delay for changing the count every second.

8.7   What are the main differences between asynchronous and synchronous interfaces in a data communication system. Explain the meaning of the term protocol, and differentiate between character and bit oriented protocols.

Determine the character transmission rate over a 1200-baud line in the synchronous serial transmission mode for a character code consisting of eight bits.

9 Interrupts

9.1 Introduction

A transfer of data from a peripheral to a microprocessor, which is microprocessor initiated, will take place when

(1) the peripheral has set a status bit that indicates the data is ready for transfer, and
(2) the microprocessor has discovered that the status bit has been set.

This implies that the microprocessor has continuously to sample the data ready line in anticipation of the status bit being set, and consequently while it is so engaged it is precluded from performing any other task.

An alternative method of initiating a transfer is to allow the peripheral to indicate directly to the processor that it is ready to transfer data by raising an interrupt. The processor can then temporarily suspend its present operations, acknowledge the interrupt and accept the transfer of the data. At the termination of the transfer the processor will return to the original program at the point of interruption and will continue with its execution. For the reasons given above it is common practice for processor manufacturers to incorporate at least one interrupt terminal on the processor chip.

9.2 The interrupt process

To initiate a transfer of data, a peripheral will set its interrupt service flip-flop whose output is connected directly to the interrupt terminal of the processor. This flip-flop stores the interrupt signal at its output until such time as the processor acknowledges it and clears the flip-flop. A simple interrupt system operating on the principle just described is shown in Figure 9.1.

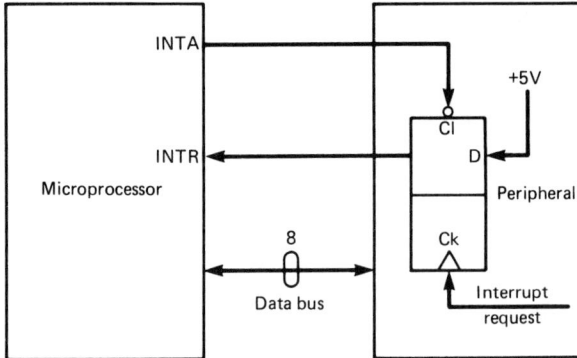

Figure 9.1 Simple interrupt system

Since the interrupt request may occur at any time it is asynchronous. On receiving the interrupt signal the processor completes the execution of the current instruction, acknowledges the interrupt and automatically saves the contents of the program counter on the stack. The processor now enters the interrupt service routine, which may be regarded as an externally initiated subroutine. If the processor registers are to be used during the course of the interrupt then instructions have to be written at the start of the service routine which will transfer the current contents of these registers to the stack. At the end of the routine the original contents of the registers and the program counter are restored and the processor returns to the execution of the main program again at the point of entry to the service routine.

9.3 Maskable and non-maskable interrupts

There are two types of interrupt in present use:

(a) the non-maskable interrupt (NMI), and
(b) the maskable interrupt (MI).

Examples of the two types of interrupt are illustrated in Figure 9.2. It will be observed that both interrupts are fed to an OR gate where they combine to generate the *master interrupt* signal. The non-maskable interrupt is fed directly to this gate while the maskable interrupt is fed via an enabling AND gate whose output is controlled by the interrupt enable flip-flop.

This flip-flop is controlled by program initiated signals. The enable interrupt signal is generated by writing the EI (enable interrupts) instruction in the program listing at some convenient point, usually the beginning of the program. In order to ensure that the execution of the main program is not interrupted, the DI (disable interrupts) instruction has to be written. This generates the reset signal for the interrupt enable flip-flop, thus disabling the AND gate in Figure 9.2.

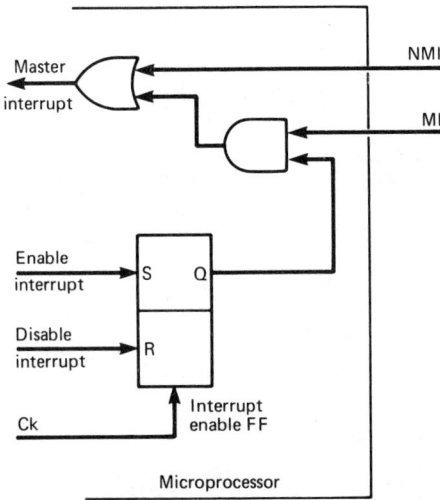

Figure 9.2 Maskable and non-maskable interrupts

In the 8085A the interrupt enable flip-flop is automatically disabled once the interrupt service routine has been entered. This ensures that a service routine in the process of being executed will not be disrupted if a second maskable interrupt is raised. At the end of the service routine the EI instruction has to be written again to re-enable the interrupt enable flip-flop.

9.4 The salient features of an interrupt system

The important features characteristic of interrupt systems are:

(1) Examination of the interrupt line by the processor.
(2) Transfer of control from the main program to the interrupt service routine.
(3) Preservation and restoration of the machine status.
(4) Identification and prioritizing of interrupt signals.
(5) Disabling and enabling of interrupts by software.

Each of these features will be discussed in more detail in the ensuing paragraphs.

9.5 Examination of the interrupt line

The general policy observed when an interrupt occurs is to complete the current instruction before transferring control to the interrupt service routine. A flowchart describing the transfer technique is given in Figure 9.3.

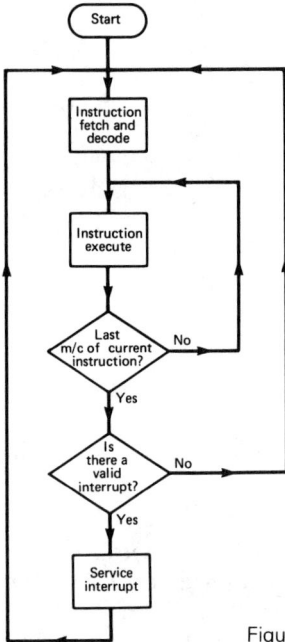

Figure 9.3 Flowchart for the examination of an interrupt

It will be observed that the processor has to make two decisions prior to servicing the interrupt. In the first instance it has to decide whether the present machine cycle is the last one associated with the current instruction, and second it must decide whether there is a valid interrupt present on the interrupt line. Assuming that the answer in both cases is in the affirmative, the processor can enter an interrupt acknowledge machine cycle, and transfer control to the interrupt service routine.

9.6 Transfer of control to the interrupt service routine

There are a number of ways in which control can be transferred from the main program to the interrupt service routine. Three of the more commonly used methods are:

(1) The TRAP, where the program counter is loaded with a known predetermined value and its original contents are transferred to the stack.

(2) Introducing a CALL instruction from external hardware. The second and third bytes of this instruction provide the starting address of the interrupt service routine. This address is initially placed in temporary storage in the processor until the current contents of the program counter have to be transferred to the stack. When this has been done, the starting address of the interrupt service routine is transferred from temporary storage to the program counter.

(3) Fetching the starting address of the interrupt service routine from a specified location in memory and inserting this address into the program counter. This memory location is accessed when the interrupt signal is received and, as in the previous case, the current contents of the program counter have been transferred to the stack.

9.7 Preservation and restoration of status

At the point of interrupt, the accumulator, the processor registers and the program counter may contain information that will be required at the end of the interrupt service routine. The information specified above is called the *status* of the machine. At the beginning of an interrupt the status has to be saved, and at the end of the interrupt the status has to be restored. The most commonly used technique for the preservation of the machine status is one in which the contents of all processor registers and the program counter are placed in temporary storage on the stack.

In the 8085A processor, raising the interrupt results in the transfer of the contents of the program counter to the stack. If it is necessary to save register contents then the initial part of the interrupt service routine will consist of stack transfer instructions. In the 6800 the contents of all registers are automatically saved at the point of interrupt.

The methods used for the preservation of status highlights one of the significant differences between the 8085A and the 6800. The 8085A has a large number of general purpose registers and may be regarded as a register oriented machine, while the 6800 has only a limited number of registers on the processor chip. Because a variable amount of program has to be reserved for transferring register contents to the stack in the 8085A, the interrupt response time is variable and if all register contents have to be saved then interrupt response is slow. In contrast, the contents of all registers are automatically saved in the 6800 and consequently it operates with a fixed response time.

At the end of an interrupt service routine, program has to be written to restore the contents of the 8085A registers to their original condition. Additionally, interrupts must be re-enabled by writing the EI instruction and this is followed by RET which, when executed, restores the contents of the program counter and returns the processor to the main program. Since the stack has a last-in, first-out structure it is possible for the processor to cope with multi-level or nested interrupts.

An alternative technique in some processors is to preserve the machine status in a second set of registers located on the processor chip. This method has the advantage of quick status retrieval; however, once the duplicate set of registers is occupied it is no longer available for servicing subsequent interrupts.

9.8 Identifying the interrupt source

When a processor is used in conjunction with a number of peripherals, one of its functions is to identify the interrupting source. Some processors have no more than one interrupt line, as in the case of the Intel 8080 and the Signetics 2650. In this case the

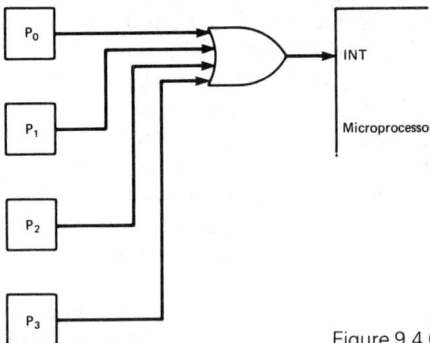

Figure 9.4 Generation of the master interrupt

various interrupts are combined in an OR gate to form the master interrupt signal which is subsequently fed to the interrupt pin of the processor (Figure 9.4).

There are two commonly used techniques available for identifying the source of the master interrupt, namely polling and vectoring. Polling is essentially a software identification method, while vectoring is the corresponding hardware approach to the same problem.

9.9 Software polling

On the receipt of the master interrupt, a processor using the polling technique described in the flowchart shown in Figure 9.5 will embark on a software polling routine. Initially, a processor register is loaded with a number m = n, where n = the number of interrupting sources. The interrupts are examined in turn, and if

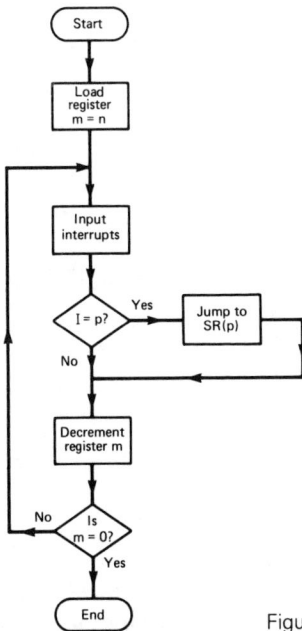

Figure 9.5 Flowchart for interrupt polling routine

the pth interrupt is raised, a jump instruction takes the processor to the starting address of the pth service routine. At the end of this routine the processor returns to the polling routine and the register is decremented. After all interrupts have been polled the contents of the register m = 0, and polling terminates.

The advantages of an interrupt polling system are that the polling routine is not initiated until the processor has received an interrupt signal, and that it is also inherently simple in that it requires no extra hardware, and is consequently cheap. However, a software overhead is involved, and as the number of interrupting peripherals increases, the method becomes inefficient because of the length of time required to poll all peripherals.

9.10 Vectored interrupts

When this method is used, the source of an interrupt is identified by means of an encoder circuit such as the one described in Chapter 2. The basic arrangement of an interrupt sorting system that utilizes an encoder is illustrated in Figure 9.6.

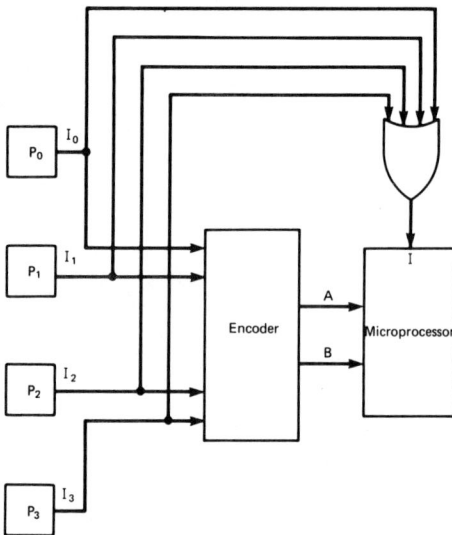

Figure 9.6 Identification of interrupting source by an encoder

The encoder detects the presence of a signal on one of its input lines and identifies the interrupting source via the address lines A and B. Simultaneously, the master interrupt signal is generated by the OR gate and is fed to the interrupt pin of the processor. In some cases an order of priority is allocated to the interrupting peripherals, and in these circumstances a priority encoder is used to identify the interrupting sources in their correct order of priority.

When several interrupting sources exist, the output of the encoder is mapped into an address which corresponds to the entry point of the interrupt service routine. An interrupt system in which the hardware supplies a separate address for each interrupting device is said to have a *vectored* interrupt structure. Such a structure provides the fastest possible interrupt servicing because time is not wasted in software polling of the peripherals; however, additional hardware is required and a hardware overhead is incurred.

In some microcomputing systems where there are a number of peripherals requiring to communicate with the processor, a Program Interrupt Controller (PIC) may be used to control the transfer of data. This device is supplied by the manufacturer as an LSI chip and has to be intially programmed with the starting addresses of the interrupt service routines of the interrupting sources. When an interrupt is raised by a peripheral and acknowledged by the processor, a CALL instruction is jammed onto the data bus by the program interrupt controller and the second and third bytes of this instruction provide the starting address of the service routine.

9.11 Hardware polling for interrupt identification

The peripherals that have to communicate with the processor are connected together in a way that allows the processor to interrogate them automatically in order to obtain the starting address of the service routine of the interrupting source. A logic diagram of such a system is given in Figure 9.7.

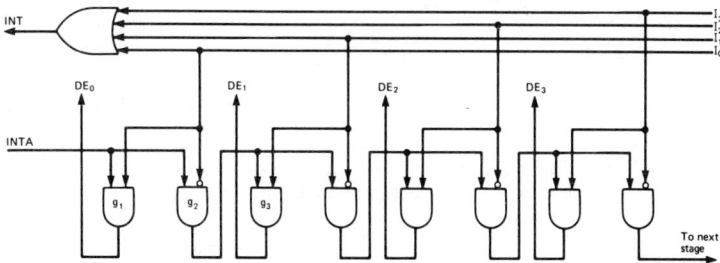

Figure 9.7 Implementation of a daisy chain system

In this system, when a peripheral raises its interrupt the master interrupt is generated in the normal way and is fed to the interrupt pin of the processor. When the processor has finished executing the current instructions, the interrupt acknowledge (INTA) signal is *daisy chained* from peripheral to peripheral. As the interrupt acknowledge signal moves down the chain it interrogates each peripheral in turn to determine which one has raised its interrupt flag. When the interrupting peripheral has been identified the acknowledge signal is blocked and prevented from proceeding further down the chain.

The peripheral now generates a data enable signal DE which activates the tri-state buffers associated with its data lines and releases a previously allocated address onto the data bus. As explained earlier, the address could be the starting address of the interrupt service routine or, alternatively, it could identify an address in memory that holds the JUMP instruction which initiates a branch to the real starting address of the interrupt service routine.

In effect, the daisy chain technique can be regarded as a hardware polling scheme. However, when the interrupting source has been recognized, a vector is released onto the data bus. Clearly, then, the daisy chain system provides a combination of polling and vectoring.

If it happens that two or more peripherals raise their interrupts simultaneously, then the interrupting peripheral closest to the start of the daisy chain is serviced first. Thus it is apparent that the system allows for the allocation of priorities to the interrupting sources.

The combinational logic required for implementing a daisy chain is very simple and is shown in Figure 9.7. Gate g_1 will provide the data enable signal DE_0 if peripheral P_0 has raised its interrupt I_0. In the absence of I_0, gate g_2 is activated and the interrupt acknowledge signal INTA is transmitted to the gate g_3, which is associated with P_1.

A hardware polling scheme such as the one described above provides a rapid method for identifying and servicing an interrupting peripheral. The only time delay involved in the system is due to the traversal of the chain of gates by the interrupt acknowledge signal, and this is very much less than the corresponding polling time in the software system.

9.12 A priority interrupt system

A typical example of a vectored interrupt is shown in Figure 9.8 where the interrupt system is capable of servicing eight interrupting peripherals. The system employs an interrupt register whose bits are set independently by the interrupt request signals from each individual interrupting peripheral. Interrupt priority is

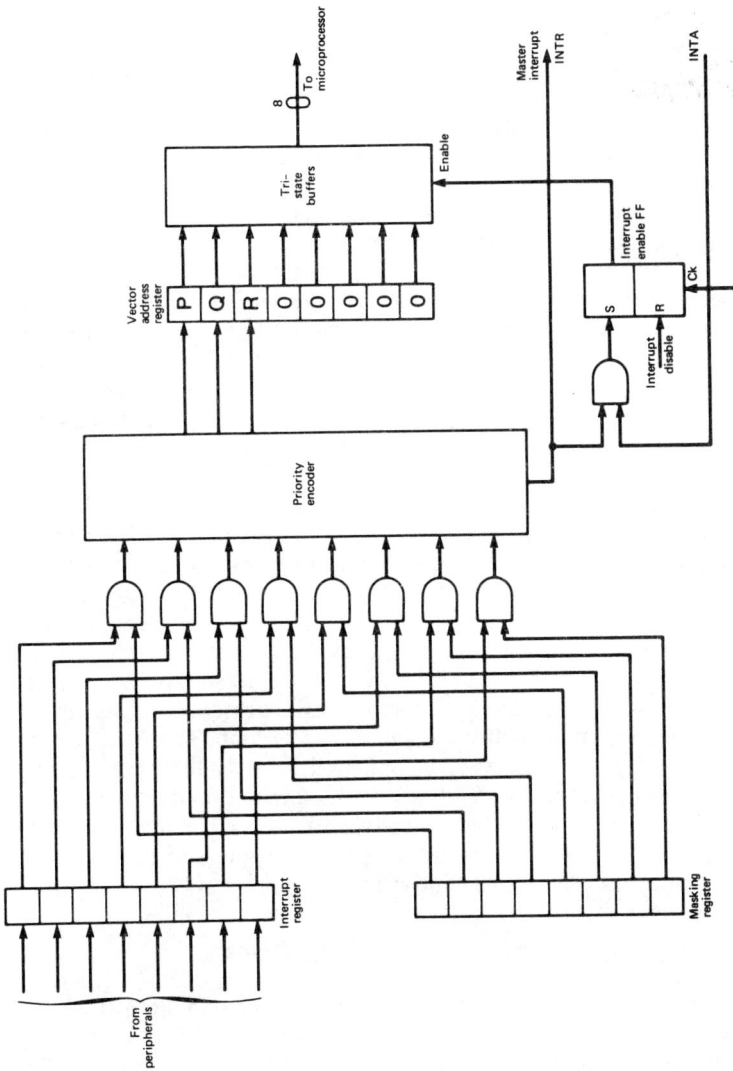

Figure 9.8 Implementation of a priority interrupt system

established by the allocation of bit positions in the interrupt register to the interrupting peripherals and also to the design of the priority encoder.

In the example shown, a programmable mask register is used which can provide an enabling or disabling signal for individual inputs. If the programmed content of a particular bit position in this register is low, then the associated AND gate is disabled, thus disabling the corresponding input. If, on the other hand, the mask bit is high, its associated AND gate is enabled and so is the corresponding interrupt signal. Clearly, an interrupt can only be recognized by the processor providing the corresponding interrupt mask bit has been set to 1 by the program.

The priority encoder generates a 3-bit vector address and a master interrupt signal. The vector address is transferred to the data bus via an 8-bit tri-state buffer and thence to the microprocessor. When the interrupt acknowledge signal INTA is returned by the processor, and providing the interrupt enable flip-flop has been set, an enable signal is generated which enables the tri-state buffers and releases the vector address onto the data bus.

The outputs of the priority encoder P, Q and R provide only three bits of the vector address and the remaining five bits can be assigned any value. In Figure 9.8 these five bits have been assigned the value 0 and consequently the vector address lies in the range 0 to 7. Additionally, the priority encoder provides its own built-in allocation of priorities to the individual interrupt inputs. For example, the encoder may have been designed so that input I_7 has the highest priority and irrespective of the presence or absence of the other seven interrupting signals the output generated will be PQR = 111. If, however, $I_7 = 0$ and $I_6 = 1$, then the priority encoder output will be PQR = 110. Assuming the priority encoder has been designed to operate in this way it is clear that its output address in the presence of interrupts corresponds to the activated interrupt with the highest subscript.

9.13 Enabling and disabling interrupts

All processors are provided with facilities for enabling or disabling interrupts. For example, the processor may be executing a program that must not be interrupted. In this case interrupts are inhibited until such time as the processor is prepared to accept them again.

In the case of the Intel 8085A, for example, two instructions are available for this purpose, namely EI, enable interrupts, and DI, disable interrupts. If the processor is prepared to accept interrupts the EI instruction is written into the program and this sets the interrupt enable flip-flop when it is executed. Alternatively, if it is not convenient to interrupt the processor, the DI instruction is written and, when executed, disables all maskable interrupts.

On entering an interrupt service routine after the interrupt acknowledge signal has been received, any further interrupts are disabled, thus preventing the interrupt from interrupting its own service routine. It is then necessary to write the EI instruction at the end of the routine in order to allow the acceptance of interrupts again when the main program has been entered.

Additionally, the interrupt system is disabled automatically after a reset, and this allows the programmer to load registers and initialize variables that may be required during interrupt service and recognition.

9.14 The 8085A interrupts

The 8085A has five interrupt pins, which are labelled TRAP, RST 7.5, RST 6.5, RST 5.5 and INT. The TRAP interrupt has highest priority and it is also a non-maskable interrupt. It causes the processor to store the contents of the program counter on the stack and it jumps to memory location 24H, which is the starting address of the TRAP routine. This interrupt could be, for example, dedicated to sensing a mains voltage failure, and the service routine associated with the interrupt could be written to save the status of the machine or, alternatively, switch to a standby supply.

The three interrupt pins RST 7.5, RST 6.5 and RST 5.5 are effectively hardware restarts. The processor response to any one of these interrupts is to store the contents of the program counter on the stack and jump automatically to one of three locations, 3CH, 34H or 2CH, each one corresponding to the interrupts in the order given above. Between RST 5.5 and RST 6.5 and between RST 6.5 and RST 7.5 there are seven memory locations. If the service routine for these interrupts is more than four bytes long, a jump instruction is used to branch to a new location in memory for the remainder of the service routine.

There are two processor registers associated with interrupts, referred to as mask registers. One of these registers can be written into by executing the instruction 'set interrupt mask' (SIM). When the instruction is executed, the contents of the accumulator are loaded into the mask register. The format of the register is shown in Figure 9.9(*a*) and it should be noted that the contents of this register cannot be read.

Bits 0–2 of the register set or reset the masks for RST 5.5, RST 6.5 and RST 7.5 respectively, providing the mask set enable bit (MSE) = 1. If bit 4 = 1 then the RST 7.5 internal request flip-flop will be reset irrespective of whether RST 7.5 is masked or not. The serial output data (SOD) latch can be loaded with bit 7 provided the serial output enable signal (SOE) bit 6 = 1.

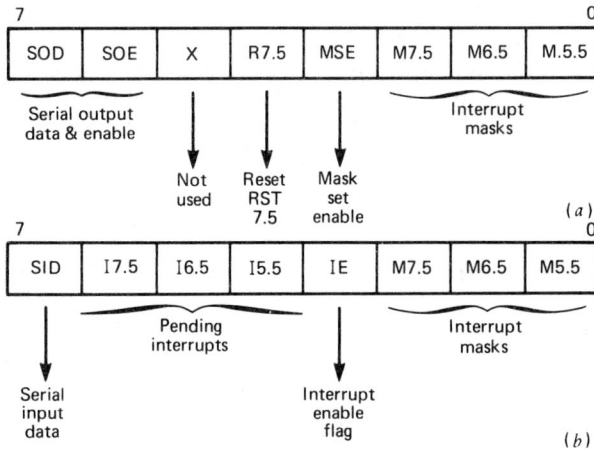

Figure 9.9 (*a*) The interrupt mask register as set by the SIM instruction (*b*) The interrupt mask as read by the RIM instruction

The following program can be used to load the mask register and enable the interrupts:

```
MVI A, 0CH    ;   Load A with masking data
SIM           ;   Set interrupt mask
EI            ;   Enable interrupts
```

When this program is executed, bits 2 and 3 will be set to 1, thus masking the interrupt RST 7.5. After the implementation of SIM, the interrupts are enabled by writing the EI instruction which, when executed, sets the interrupt enable flip-flop.

The second register can be read by writing the instruction RIM (Read Interrupt Mask). The format of this register is shown in Figure 9.9(*b*). Bits 0, 1 and 2 give the status of the masks. Bit 3 is the interrupt enable flag, and when it is set to 1 the interrupt system has been enabled by the EI instruction. Bits 4, 5 and 6 are called the interrupt pending flags. When they are set to 1 an interrupt is being requested on the corresponding RST line but has not yet been accepted. Bit 7 simply gives the contents of the serial input data line. After the execution of the RIM instruction the contents of this register are transferred to the accumulator.

9.15 State diagram for the 8085A, including interrupt logic

In the chapter on microprocessor architecture (Chapter 4), the simplified state diagram for the 8085A was given. This diagram is reproduced in Figure 9.10 but, in this case, includes the interrupt logic. This logic first determines whether the present machine cycle is the last one for the instruction that is being presently executed, and secondly, whether a valid interrupt has been raised.

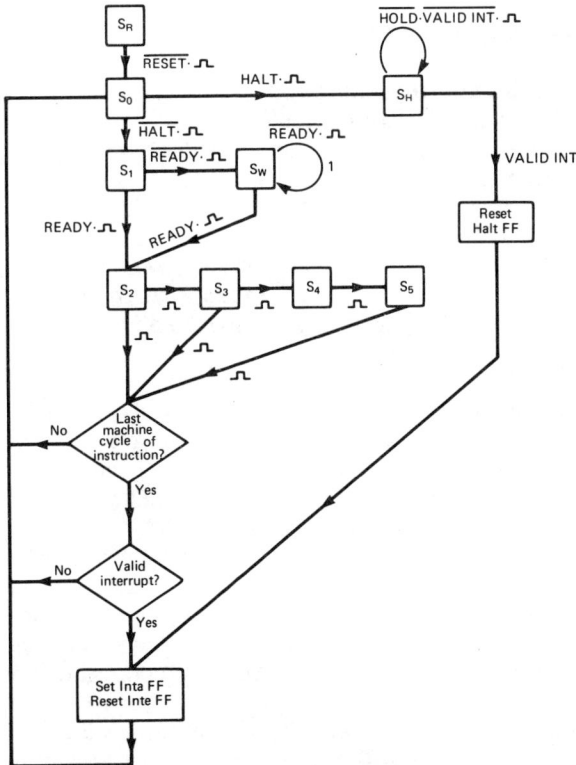

Figure 9.10 Simplified state diagram for the 8085A, including interrupt logic

Alternatively, if the machine happens to be in the HALT state S_H, the interrupts are examined during every clock cycle. If a valid interrupt signal is present the interrupt acknowledge flip-flop is set and the interrupt enable flip-flop is reset. The next machine cycle is a special one for handling interrupts. If the interrupt appears on TRAP or any of the RST lines, the machine enters a BUS IDLE cycle; if it is INTR the machine enters an INA (interrupt acknowledge cycle).

The condition for a valid interrupt in the 8085A is given by the following Boolean expression:

$$\text{VALID INT} = \text{ML}[\text{TRAP} + \text{EI}(\text{INTR} + \text{RST } 7.5 \,.\, \overline{\text{M7.5}} + \text{RST } 6.5 \,.\, \overline{\text{M6.5}} + \text{RST } 5.5 \,.\, \overline{\text{M5.5}})]$$

where ML = 1 if the processor has entered the last machine cycle of the current instruction.

9.16 Interrupt response of the 8085A

The CPU receives an interrupt on the INTR (interrupt request) pin, providing the INTE (interrupt enable) flip-flop has been set by the enable interrupt (EI) instruction. Besides sampling the INTR pin, the processor also tests the status of the TRAP and RST pins during a clock cycle just prior to the termination of the present instruction. If INTR is the only valid interrupt, and if the interrupt enable flip-flop is set, the processor enters an INA cycle. This cycle is identical to an opcode fetch, except that an interrupt acknowledge signal ($\overline{\text{INTA}}$) is generated rather than a read ($\overline{\text{RD}}$) signal.

When $\overline{\text{INTA}}$ is generated, external hardware provides the opcode of an instruction to execute. The opcode is placed on the data bus and is read by the processor. If it is a multi-byte instruction then additional INA machine cycles are generated to transfer the additional bytes to the processor. The external hardware may provide either a CALL or an RST instruction. Either of these instructions is suitable because they both transfer the present contents of the program counter onto the stack. If the external hardware provides the CALL instruction, then there are three bytes to this instruction; the first is the opcode while the succeeding two bytes provide the 16-bit starting address of the interrupt service routine.

After receiving the opcode and decoding it, the processor knows that the CALL instruction consists of three bytes. It then performs two more memory read machine cycles to bring the starting address of the interrupt service routine to temporary storage and, at the same time, incrementation of the program counter is inhibited. Having received the three bytes of the instruction, the processor executes it by transferring the contents of the program counter to the stack, and the starting address to the program counter. This has the effect of jumping the execution of the program to the memory location specified in the second and third bytes of the instruction. Two memory write machine cycles are used for this transfer; therefore, in all, an interrupt acknowledge cycle consists of five distinct machine cycles, three for reading the CALL instruction into the processor, and the last two for writing the contents of the program counter onto the stack.

Alternatively, external hardware could supply the RST instruction, a one-byte instruction, whose bit pattern is

1 1 X X X 1 1 1

where XXX is a 3-bit binary number in the range 000 to 111. This bit pattern can be released on to the system data bus by the $\overline{\text{INTA}}$ signal as illustrated in Figure 9.11. When the instruction decoder

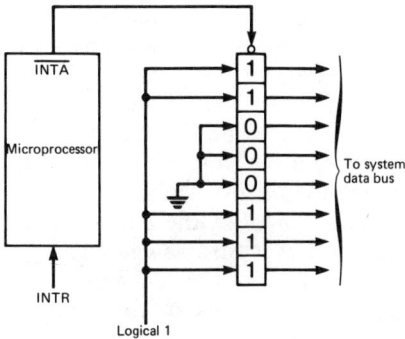

Figure 9.11 Release of RST vector onto the system data bus

interprets this instruction and the processor executes it, the following three events occur:

(1) the program counter is inhibited,
(2) the contents of the program counter are transferred to the stack, and
(3) are replaced automatically by the bit pattern 0000000000XXX000, i.e. by one of eight values depending upon the bit pattern of the RST instruction. Hence the program counter may contain any one of the following addresses, 00H, 08H, 10H, 18H, 20H, 28H, 30H and 38H.

The bit pattern of the RST instruction is referred to as the interrupt vector, and the preliminary starting address of the interrupt service routine is given by one of the eight addresses above. There are only eight memory locations separating each of these addresses, which is sufficient to store the status of the machine on the stack but not the complete interrupt service routine. A JUMP instruction is required which, when implemented, will provide a further address which is the starting address of the remainder of the interrupt service routine.

For TRAP, RST 7.5, RST 6.5 and RST 5.5, control is transferred to the interrupt service routine by the mechanism just described, except that in this case the RST instruction is internally generated by the microprocessor hardware on receipt of the interrupt. The processor now enters a BUS IDLE cycle in which the program counter is inhibited, and its contents are transferred to the stack. The starting addresses of the interrupt service routines associated with these four interrupts have already been specified in Section 9.14.

In all cases, if the entire status of the machine is to be saved, the contents of all registers have to be transferred to the stack at the beginning of the interrupt service routine. This is done by writing the PUSH instruction for four consecutive memory locations at the beginning of the service routine to cover the transfer of the contents of the B, D, H and PSW registers to the stack.

At the end of the interrupt service routine the status of the machine is restored by the execution of the POP instruction four times and in the reverse order to the PUSH instructions previously specified. The POP instructions are followed by the EI instruction which allows the processor to respond to interrupts again and the RET instruction which returns the address of the next instruction in the main program sequence from the stack to the program counter.

9.17 The 6800 interrupt structure

There are three hardware interrupts associated with the 6800. They are reset ($\overline{\text{RES}}$), non-maskable interrupt ($\overline{\text{NMI}}$) and interrupt request ($\overline{\text{IRQ}}$). A program interrupt can be initiated by supplying an active low signal to the $\overline{\text{IRQ}}$ pin on the processor. This signal would normally be generated by a PIA (peripheral interface adaptor) or by an ACIA (asynchronous control interface adaptor). To prevent disruption of the current activity of the machine on the receipt of an interrupt, the whole of the machine status is transferred to the stack so that, after processing the interrupt, the machine is returned to its pre-interrupt condition almost immediately.

A flow chart for the interrupt sequence is shown in Figure 9.12. At the end of the present instruction the status of the interrupt mask is checked. If $I_m = 1$, interrupts are inhibited and the present program sequence continues; if $I_m = 0$, the contents of the processor registers are stored on the stack.

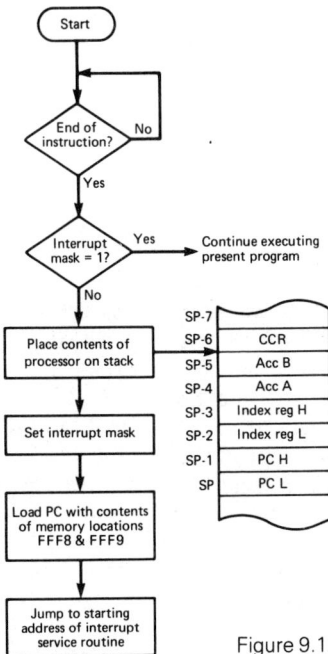

Figure 9.12 Flowchart for an interrupt request sequence

If any further interrupts are to be inhibited while the present interrupt is being serviced the interrupt mask I_m is set to 1. It is then cleared at the end of the interrupt service routine by writing the return from interrupt (RTI) instruction. Alternatively, if a multi-level interrupt system is to be established, the interrupt mask can be cleared at the start of the present interrupt service routine by writing the CLI instruction, thus allowing a second interrupt to be serviced before the first has been cleared.

After setting the interrupt mask, the processor accesses memory locations FFF8H and FFF9H where the starting address of the interrupt service routine is found. This technique of indirect addressing is also used by the other interrupt signals, and the interrupt vectors are placed in memory locations FFF8H to FFFFH, as shown in Figure 9.13.

RES (low byte)	FFFF
RES (high byte)	FFFE
NMI (low byte)	FFFD
NMI (high byte)	FFFC
SWI (low byte)	FFFB
SWI (high byte)	FFFA
IRQ (low byte)	FFF9
IRQ (high byte)	FFF8

Figure 9.13 Memory assignment of interrupt vectors

The non-maskable interrupt, as its name suggests, is the one with the highest priority. When an interrupt signal appears on the NMI pin it is serviced immediately. The starting address of the NMI service routine is found in memory locations FFFCH and FFFDH, and the contents of these locations are transferred to the program counter. Like the TRAP interrupt in the 8085A, NMI is used in the event of a power failure, or, alternatively, for initiating the service routine for a peripheral that must always be allowed to interrupt.

The RES interrupt is used when the power is initially switched on, and it initiates the service routine for setting up the initial values of the program counter and the stack pointer. The starting address of the routine is found in memory locations FFFFH and FFFEH, and in this case there is no requirement to store the processor registers.

If there are a number of peripherals associated with the system, individual interrupts can be connected via an OR gate to the IRQ terminal of the processor, thus generating the master interrupt signal. The interrupt signals generated by the peripherals may be routed through a PIA, where, in addition to generating the master interrupt, they are used to set status flags. The processor then polls the status flags and priority is established by the order in which they are polled.

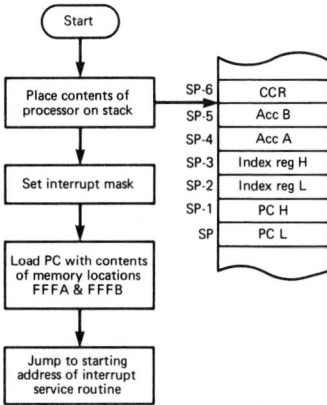

Figure 9.14 Flowchart for a system interrupt

An interrupt can also be initiated by writing the SWI (software interrupt instruction). A flowchart illustrating the interrupt sequence is shown in Figure 9.14. It is similar to the hardware interrupt sequences, and the starting address of the interrupt service routine is obtained from memory locations FFFAH and FFFBH after the contents of the processor register have been stored.

9.18 Direct memory access (DMA)

For the normal program controlled transfer of data from a peripheral to memory, the transfer path is as shown in Figure 9.15. The data would be transferred from peripheral to input port, from there to the accumulator and thence to memory.

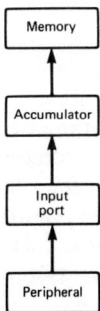

Figure 9.15 The normal input path for a microprocessor

A simple program for transferring one byte of data from an input port to a specified memory location in an 8085A system would be:

```
IN      01H    ; Input data
STA     4800H  ; Store data
```

The execution of these two instructions would require 23 clock cycles and the transfer would take approximately 7.5 μs. In practice, the transfer of one byte out of a block of data will take considerably longer than 7.5 μs since the process will involve the checking of a status bit, the incrementation of a memory pointer, and the decrementation of a counter. A transfer time of 20 μs would not therefore be unreasonable and this is equivalent to a data transfer rate of 50 kilobytes/sec.

A typical memory element may well have an access time of the order of 1 μs and some peripherals, for example a disk store, have data transfer rates much in excess of 50 kilobytes/sec. For this reason transfers directly between peripheral and memory are frequently employed using the DMA path, shown in Figure 9.16, in order to provide faster data transfers than are possible through the normal transfer path of Figure 9.15.

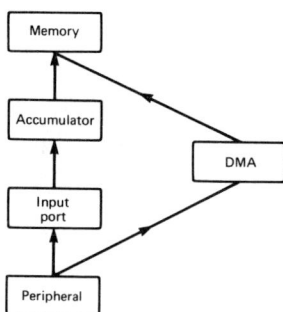

Figure 9.16 The DMA path for data transfer

The idealized diagram of Figure 9.16 shows a separate channel for DMA but, in practice, interconnections in a microprocessor system are via the system bus, which consists of three components, data, address and control. It is therefore desirable to share the bus system, as shown in Figure 9.17. Such a scheme requires that either the microprocessor or the DMA controller will have control of the bus system, and the DMA controller may be regarded as a second processor which takes over control of the bus system during a DMA transfer.

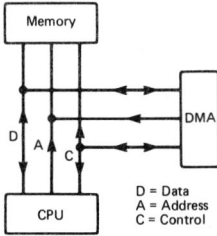

Figure 9.17 Bus sharing by CPU and DMA

9.19 The HOLD state

Processors with a DMA capability have a terminal such as the HOLD terminal on the 8085A chip, which is usually activated by a peripheral via a DMA controller when a DMA transfer is required.

In the simplest case of a three-state machine cycle, the 8085A samples the HOLD request pin during state S_1, as illustrated in Figure 9.18. This state immediately follows the address presentation state S_0. If the HOLD pin is high when sampled, the HLDAFF (Hold Acknowledge flip-flop) is set, the present machine cycle is terminated and if now HLDAFF = 1, the processor enters the HOLD state and releases control of the bus system. The DMA transfer now takes place under the control of a DMA controller and is eventually terminated when the HOLD pin goes low. At this point the processor regains control of the bus system and enters the next machine cycle.

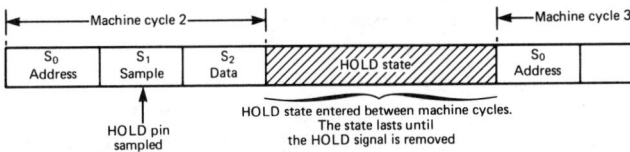

Figure 9.18 Entry into the HOLD state

The state diagram of the 8085A is repeated again in Figure 9.19, on this occasion showing the DMA logic only. It will be observed that the HOLD state S_{HLD} can be entered at the termination of a 3-, 4- or 6- state machine cycle as described above, or, alternatively, from the HALT state S_H.

If the processor is in the HALT state and the HOLD pin goes high, the Hold Acknowledge FF is set and the HOLD state S_{HLD} is entered. Providing the HOLD signal is maintained, the processor

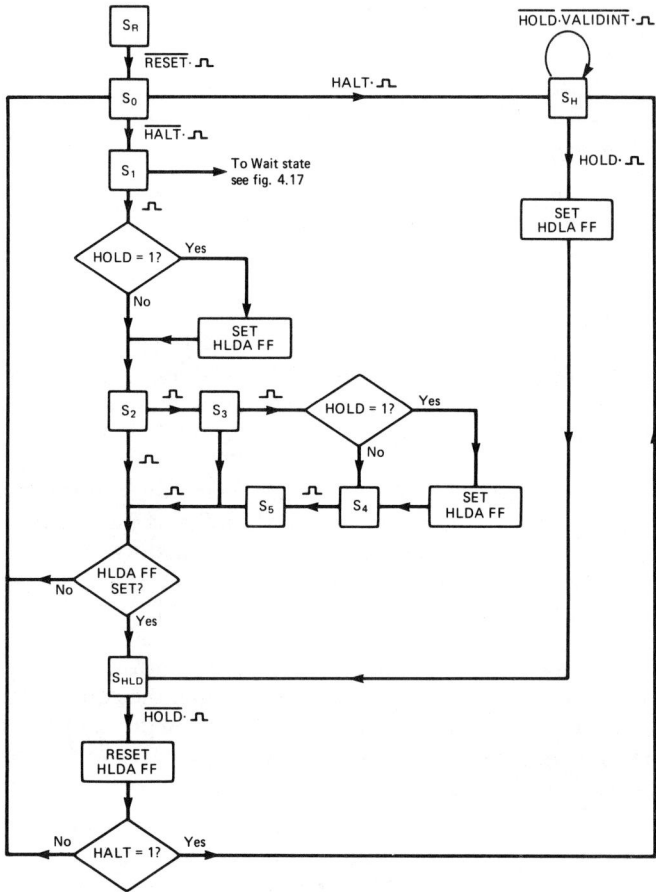

Figure 9.19 Simplified state diagram for the 8085A, including DMA logic

remains in this state. On the first clock pulse after the HOLD signal has gone low, the processor either enters the next machine cycle or, if the HALT signal = 1, it returns to the HALT state. At the same time the Hold Acknowledge FF is reset.

9.20 Basic requirements of a DMA transfer

The block diagram of Figure 9.20 shows the basic interconnections for a DMA transfer. A DMA Request is raised by the peripheral and is transmitted to the DMA controller which, in turn, generates

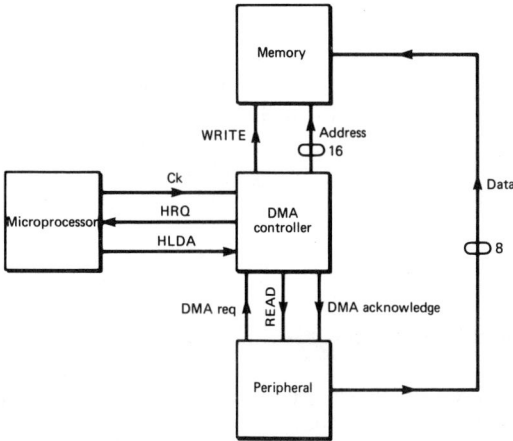

Figure 9.20 Basic interconnections of a DMA system

a Hold Request (HRQ) signal. This HRQ signal is connected to the HOLD pin of the processor. After the processor has sampled the HOLD pin, the Hold Acknowledge (HLDA) signal is returned to the DMA controller, and the processor relinquishes control of the bus system. The DMA controller now takes over the bus system, generates a DMA Acknowledge signal, which is returned to the peripheral, and produces a READ signal which releases data from the peripheral onto the data bus. When the data is stable on the data bus, the WRITE signal is generated by the controller and the data on the data bus is written into the addressed memory location.

9.21 Basic requirements of a DMA controller

The basic requirements of a DMA controller are:

(1) An address register, which contains the address in memory where the first data byte is to be either stored or read. This register is incremented as each byte of data is transferred. Additionally, the register has to be initialized when the controller is operating in the slave mode.

(2) A count register, which contains the length of the block of data to be transferred. This register is then decremented as each byte of data is transferred. This register also has to be initialized when the controller is operating in the slave mode.

The controller will normally have three modes of operation. They are:

(1) The master mode, when the DMA controller is in charge of the bus system and a DMA transfer is taking place.
(2) The slave, or initializing mode, when the contents of the address and counter register are programmed, and
(3) The non-operating mode, when it is in neither the master nor the slave mode.

9.22 DMA with multiple external devices

In practice, one DMA controller can control more than one peripheral. Since the DMA controller does not actually transfer any data itself, it does not require any I/O ports through which the peripherals can communicate with memory. The peripherals connect directly to the data bus and a transfer is made when the peripheral is instructed to do so by the DMA controller.

Since, in this case, the DMA controller is controlling four peripherals, it must contain four separate address registers and four separate count registers. The controller may also contain prioritizing circuits for deciding which of the peripherals has highest priority if more than one DMA request signal is raised simultaneously. Clearly, the DMA controller will contain a significant amount of logic, comparable with that of the microprocessor itself.

9.23 The Intel 8257 DMA controller

This chip, when used in conjunction with the 8212 8-bit latch, can provide a DMA capability for four peripherals. A block diagram illustrating the interconnections between the 8257 and the 8212, and the external connections to and from the 8257, are shown in Figure 9.21. The 8257 contains four 16-bit address registers and four 14-bit count registers. After initialization, the controller can control the transfer of a block of data of maximum length 16 384 bytes for each of the four channels.

The controller enters the slave mode when the $\overline{CS}$ line is low, while to enter the master mode the controller must receive the HLDA signal from the 8085A microprocessor.

DRQ_0–DRQ_3, the DMA request inputs, are used by the peripherals to obtain a DMA cycle. DRQ_0 has the highest priority

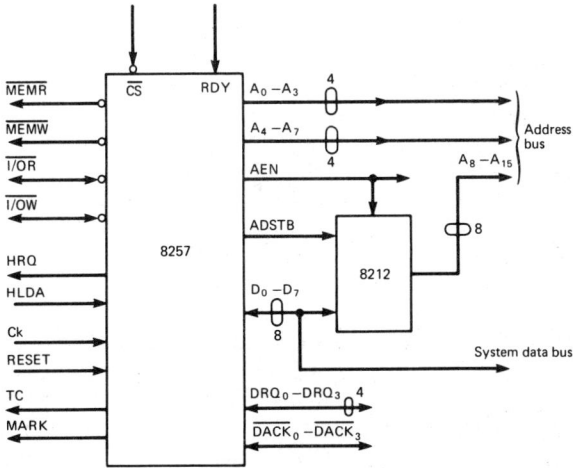

Figure 9.21 Interconnections for the 8257 DMA controller

of these four request signals unless the rotating priority option is being employed. The DRQ signal from the selected peripheral will remain high until the termination of the DMA transfer.

$\overline{DACK}_0$–$\overline{DACK}_3$ are the DMA acknowledge outputs of the controller, generated when it is in receipt of the HLDA signal from the 8085A. The appropriate DACK signal is generated to select the required peripheral and, in effect, acts as a chip select.

The data lines D_0–D_7 are bidirectional and serve two functions. When the controller is in the slave mode they carry the data from the 8085A that is used to initialize the address and count registers, while in the master mode they output the eight most significant address bits to the 8212 latch. These address bits are strobed into the latch by the address strobe (ADSTB) and they are released onto the address bus by the address enable (AEN) signal which is additionally used to disable the system data and control buses.

Address lines A_0–A_3 are also bidirectional. In the slave mode they are the inputs that select the DMA registers to be initialized, while in the master mode they carry the four least significant address bits. Thus, when a DMA transfer takes place the memory address is output on lines D_0–D_7 and A_0–A_7.

A terminal count signal TC is generated by the controller and is used to notify the currently selected peripheral that the byte presently being transferred is the last one for this data block. The MARK output notifies the selected peripheral that the current DMA cycle is the 128th cycle since the previous MARK output.

A single byte of data can be transferred using the 8257 controller, in four clock periods. For the 8085A, the clock period is 320 ns, hence the transfer time for 1 byte is 1.28 μs, which is equivalent to a transfer rate of 781 kilobytes/sec., a faster transfer rate than is possible when using the conventional I/O path. If it is necessary to increase the DMA transfer time because it is too short for the memory chip being used, then a READY signal is provided by the memory chip which can elongate the transfer time by an integral number of clock periods. If the READY signal is low, the controller moves into a WAIT state where it remains until such time as the READ signal and clock are simultaneously high.

The $\overline{\text{MEMR}}$ and $\overline{\text{MEMW}}$ strobes are generated by the controller when data is either to be read from the addressed memory location or, in the case of $\overline{\text{MEMW}}$, when it is to be written into the addressed memory location. Similarly, when operating in the master mode $\overline{\text{I/OR}}$ and $\overline{\text{I/OW}}$ are strobes generated by the controller either to access data from, or write data into, the selected peripheral. These two lines are bidirectional, and when the controller is operating in the slave mode they carry input signals which allow the address and counter registers to be either read or loaded.

Problems

9.1   Write a program which, in response to an interrupt request signal, polls four peripherals and services them in sequence. The interrupt service routines are located at memory locations identified by the labels PER 1, PER 2, etc. Prior to responding to an interrupt request signal the state of the machine has to be saved.

9.2   A main program sequence consists of incrementing continuously the contents of the program counter. This program is interrupted by an external peripheral to which a block of data, whose starting address is 0100H, is to be transferred. Write an assembly language program for both the main program sequence and the interrupt service routine, and return to the main program sequence at the end of the transfer of data.

9.3   People entering a museum are counted by an electronic digital system. The count has to be registered in the memory of a microcomputing system based on the 8085A by incrementing the contents stored in two adjacent memory locations previously specified for this purpose. The count is incremented by jamming an interrupt vector onto the data bus by external hardware. Develop a suitable hardware system using the 8085A INTR pin, and write a suitable interrupt service routine in assembly language.

9.4   Six I/O devices are to communicate with the 8085A via its interrupt pin. Develop a logic system which provides an interface between the I/O devices and the microprocessor using a priority encoder and any other logic elements needed.

The output of the priority encoder forms part of a vector which is used as an index to a jump table. Write an assembly language program that facilitates the transfer of control to service routines which are associated with each of the peripherals.

9.5   An example of a daisy chain priority interrupt system which releases a different vector address on to the data bus for each stage of the chain is illustrated in Figure 9.7. Develop the logic required for each stage in the chain and draw the appropriate logic diagram.

What are the advantages of a priority interrupt system in comparison with a non-priority system?

9.6   Draw a block diagram indicating the position of a DMA controller in a microcomputer system and show the various interconnections that are required for a DMA transfer. Indicate the function of each interconnection and discuss the various logical elements you might expect to find in a DMA controller.

Draw a flowchart for describing the transfer of data from peripheral to memory, or vice versa.

10 Data conversion

10.1 Introduction

The real world in which the microprocessor, and, for that matter, the digital engineer operates, is an analogue world. However, the methods now commonly used for processing data are predominantly digital. In order to use digital processing techniques, the microprocessor engineer has to develop an interface between the analogue world and the digital machine. This interface function is provided by a device generally referred to as a *data converter*.

There are two types of data converter in general use. The analogue-to-digital converter (ADC) is used to transform analogue data into digital data. When the digital processing has been performed in the digital machine, the result may be stored in memory or be displayed on a VDU, for example, or alternatively, it may be transformed again into analogue form by a digital-to-analogue converter (DAC). The transformed data can then be used to control the behaviour of an analogue device or process.

The general principles of both types of converter will be studied in this chapter. However, choosing a data converter for a practical application requires a careful examination of the manufacturer's data sheet, but nevertheless, the contents of this chapter should enable the reader to make an intelligent choice.

10.2 Basic principles of digital-to-analogue conversion

The output voltage of an operational amplifier circuit such as the one illustrated in Figure 10.1 is given by the equation

$$V_o = - I_f R_f$$

since the point X on the diagram can be regarded as a *virtual earth*.

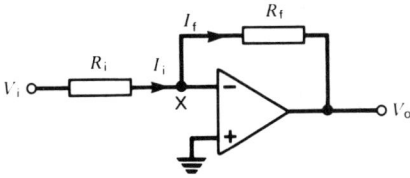

Figure 10.1 Basic operational amplifier circuit

The input impedance of this amplifier is very high, consequently

$$I_i = I_f$$

Hence the input voltage V_i is given by

$$V_i = I_i R_i = I_f R_i$$

and the gain of the amplifier

$$A = -\frac{V_o}{V_i} = -\frac{R_f}{R_i}$$

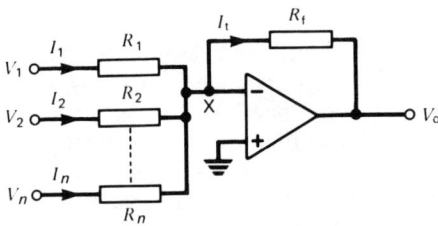

Figure 10.2 The summing and inverting amplifier

By a simple modification, the basic operational amplifier can be converted to a summing and inverting amplifier as shown in Figure 10.2, where

$$I_t = (I_1 + I_2 + \ldots I_n)$$

and

$$I_1 = \frac{V_1}{R_1} ; \quad I_2 = \frac{V_2}{R_2} ; \quad \ldots I_n = \frac{V_n}{R_n}$$

The output voltage V_o is given by

$$V_o = -I_t R_f$$
$$= -\left(\frac{V_1}{R_1} + \frac{V_2}{R_2} \ldots \frac{V_n}{R_n} \right) R_f$$

If all the inputs to this circuit are connected to a common *reference voltage* V_r the preceding equation may be written

$$V_o = -V_R \left(\frac{R_f}{R_1} + \frac{R_f}{R_2} + \ldots \frac{R_f}{R_n} \right)$$

Additionally, if the circuit resistors are weighted so that

$$R_f = R; \quad R_1 = 2R; \quad R_2 = 4R; \quad \text{and } R_n = 2^n R$$

then

$$V_o = -V_R \{2^{-1} + 2^{-2} + \ldots 2^{-n}\}$$

Assuming that facilities are now provided so that individual resistors can be switched in and out of the circuit at will, the equation may be modified to read

$$V_o = -V_R(B_1 2^{-1} + B_2 2^{-2} + \ldots B_n 2^{-n})$$

where $B_1 = B_2 = \ldots B_n = 0$ indicates the absence of a resistor from the circuit and $B_1 = B_2 = \ldots B_n = 1$ indicates the presence of a resistor in the circuit.

The section of the above equation enclosed in brackets represents a fractional binary series, and the presence or absence of any given term in the series depends upon whether the value of its associated binary coefficient B is either 1 or 0.

As an example, consider the case of a 3-bit conversion with $V_R = -10\,\text{V}$. The fractional binary series is then limited to three terms, and if the coefficients associated with these terms are $B_1 = 1$, $B_2 = 0$ and $B_3 = 0$, then

$$V_0 = 10 \,(1 \times 2^{-1} + 0 \times 2^{-2} + 0 \times 2^{-3})$$
$$= 5\,\text{V}$$

and the output voltage is equal to half the reference voltage V_R.

Similarly, for $B_1 B_2 B_3 = 010$, $V_o = 2.5\,\text{V}$; and for $B_1 B_2 B_3 = 110$, $V_o = 7.5\,\text{V}$.

The maximum voltage appearing at the output occurs when the binary code at the input is $B_1 B_2 B_3 = 111$ and is $V_o = 8.75\,\text{V}$. For 3-bit conversion there are eight possible input codes and these give rise to eight different analogue output voltages, as shown in Figure 10.3(*a*). This table also includes a column for the *normalized* output voltage V_o/V_R which is plotted against the binary input code in Figure 10.3(*b*). No other analogue output voltages are available except those shown in the tabulation, and consequently a bar graph of the eight discrete values in the table is obtained.

If 001 is added to the input code word 100, a new input code of 101 is obtained and the analogue output voltage changes by a

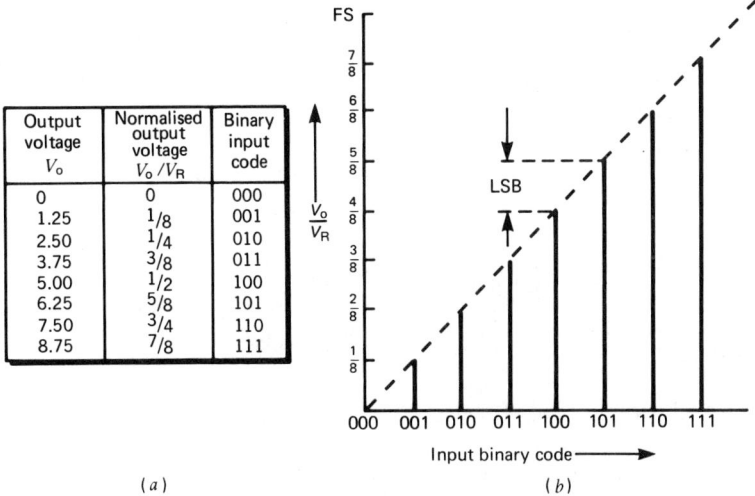

Output voltage V_o	Normalised output voltage V_o/V_R	Binary input code
0	0	000
1.25	$1/8$	001
2.50	$1/4$	010
3.75	$3/8$	011
5.00	$1/2$	100
6.25	$5/8$	101
7.50	$3/4$	110
8.75	$7/8$	111

(a)   (b)

Figure 10.3 (a) Tabulation of output voltage and binary input code for a 3-bit DAC
(b) Graph of normalized output voltage against input binary code for a 3-bit DAC

discrete amount equivalent to the least significant bit (LSB) as indicated on Figure 10.3(b) and equal to 1.25 V. This discrete voltage step is called the *step size*.

It is worth noting that in the case of the 3-bit converter the maximum output voltage $V_{OM} = 8.75$ V and is well below the reference voltage. However, as the number of digits in the input binary code is increased V_{OM} approaches the reference voltage and becomes nearly equal to it for a large number of input bits.

10.3 The weighted resistor DAC

The circuit of a weighted resistor DAC is shown in Figure 10.4. The weighted resistors are connected to virtual earth X through the switch contacts marked 1 or, alternatively, to ground via the contacts marked 0. If the current I is defined by the equation $I = V_R/R$ and assuming that all the coefficient switches are in the 1 position, then

$$I_t = V_R/2R + V_R/4R + V_R/8R$$
$$= I (2^{-1} + 2^{-2} + 2^{-3})$$
$$= 7/8I = (1 - 2^{-3}) I$$

In the general case for n resistors

$$I_t = I(1 - 2^{-n})$$

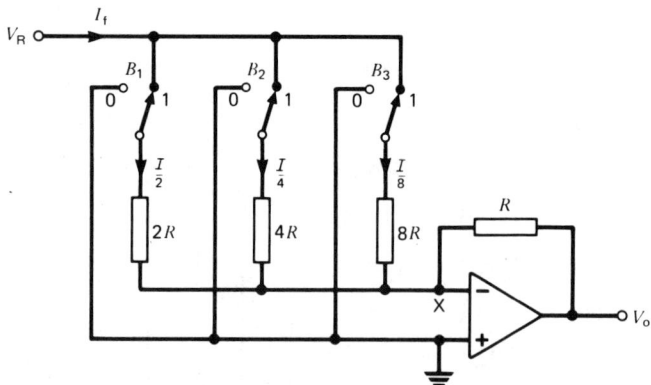

Figure 10.4 The weighted resistor DAC

The maximum output of the converter

$$V_o = -I_t R$$

hence

$$V_o = -V_R\{B_1 2^{-1} + B_2 2^{-2} + B_3 2^{-3}\}$$

The coefficient switches in Figure 10.4 are shown as electromechanical devices, but in practice they are electronic analogue switches. A typical example of such an electronic switch employing MOSFETs is illustrated in Figure 10.5. In this circuit a totem pole MOSFET driver is connected to one end of the DAC input resistors, while the other end of these resistors is connected to the operational amplifier input line. Binary data in both true and

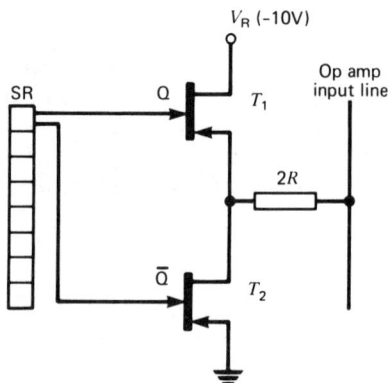

Figure 10.5 The MOSFET analogue switch

complemented form Q and $\bar{Q}$ is available from a shift register and is connected to the gates of transistors T_1 and T_2.

If $Q = 1$ and $\bar{Q} = 0$, T_1 is turned ON while T_2 is OFF, thus connecting V_R directly to the left-hand end of the DAC input resistor. Conversely, if $Q = 0$ and $\bar{Q} = 1$, T_1 is OFF and T_2 is ON, thus connecting the left-hand end of the DAC input resistor directly to ground.

The accuracy of the converter (Figure 10.4) depends upon the accuracy of the resistors. The resistance values in the network are all different and the largest value in the chain is given by $2^n R$, where n is the number of bits in the input code. For an 8-bit converter in which $R = 5\,\mathrm{k\Omega}$, the largest resistance in the chain will have a value of $2^8 \times 5\,\mathrm{k\Omega} = 1.28\,\mathrm{M\Omega}$ and for a 12-bit converter the largest resistance value is $2^{12} \times 5\,\mathrm{k\Omega} = 20.48\,\mathrm{M\Omega}$. Such large resistance values are difficult to achieve on integrated circuits and this has led to the development of the *R/2R* ladder network DAC.

10.4 the use of *R/2R* ladder networks in DACs

The difficulty of providing a wide range of precision binary weighted resistors is overcome in many practical converters by the use of an *R/2R* ladder network. This type of network employs two values of resistance only, R and $2R$, while providing binary weighted bit currents. Since only two resistance values are required in the circuit the process of trimming the resistors in manufacture is greatly simplified.

A circuit for a 3-bit *R/2R* converter is given in Figure 10.6(*a*) and it will be observed that the circuit current flows to ground when the switches are in the 0 position and to the virtual earth X when in the 1 position.

Looking to the right of terminals A and B, the input resistance is $2R\,\Omega$ (Figure 10.6(*b*)) and, similarly, looking to the right at both terminals C and D (Figure 10.6(*c*)) and E and F (Figure 10.6(*d*)) the input resistance is also $2R\,\Omega$. The equivalent circuit of the whole ladder network appears in Figure 10.6(*e*) and the total current taken from the source V_R is given by $I = V_R/R$.

At each of the nodes E, C, and A the current divides into two equal halves so that the current through the resistance of $2R\,\Omega$ between E and F is $I/2$ and the current through the resistance R between E and C is also $I/2$. Now

$$V_R = V_E = I/2 \times 2R = IR$$
$$V_C = I/4 \times 2R = IR/2 = V_R/2$$
$$V_A = I/8 \times 2R = IR/4 = V_R/4$$

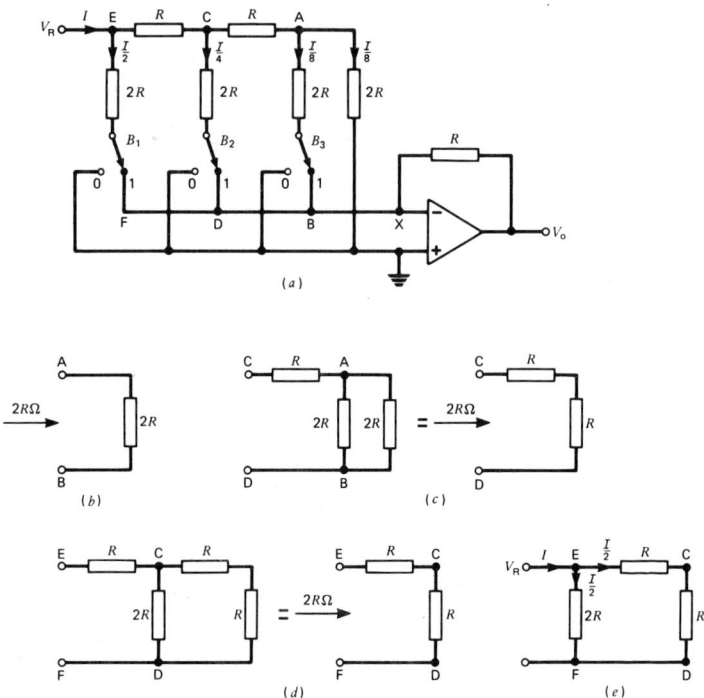

Figure 10.6 (a) The R/2R DAC circuit. Equivalent circuits looking in at: (b) terminals A and B (c) terminals C and D (d) terminals E and F (e) Equivalent circuit of the whole ladder network

Hence

$$V_o = -(V_R B_1 R/2R + V_R/2 B_2 R/2R + V_R/4 B_3 R/2R)$$
$$= -V_R (B_1 2^{-1} + B_2 2^{-2} + B_3 2^{-3})$$

which is identical to the equation for the weighted resistor DAC.

10.5 Multiplying D/A converters

The digital inputs to the DACs previously described have fixed levels of 0 and 5 volts. However, the reference voltage V_R can be of either positive or negative polarity since the analogue switches can be regarded as a bar of semiconducting material which will allow current flow in either direction. If the reference voltage is bipolar and variable, the DAC can be converted into a device

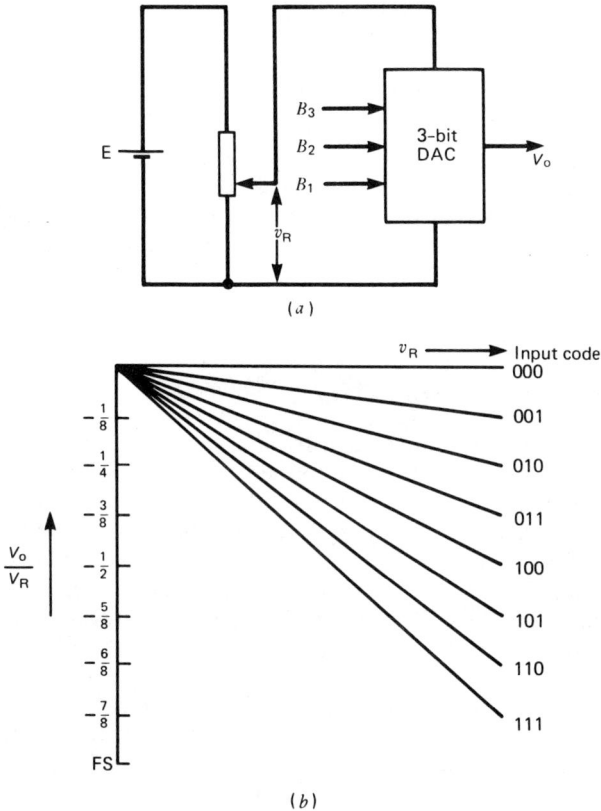

(a)

(b)

Figure 10.7 Single quadrant multiplying DAC (a) Block diagram (b) Characteristics

which performs the process of multiplication. Such a device is called a multiplying D/A converter.

If the 3-bit DAC whose block diagram is given in Figure 10.7(a) has a variable positive reference supply, the equation for its output voltage is

$$v_o = -v_R (B_1 2^{-1} + B_2 2^{-2} + B_3 2^{-3})$$

but the equation for a fractional 3-bit binary number is

$$(N_f)_{10} = B_1 \times 2^{-1} + B_2 \times 2^{-2} + B_3 \times 2^{-3}$$

and for an input code $B_1 = B_2 = B_3 = 1$, $(N_f)_{10} = 7/8$
Hence

$$v_o = -7/8 v_R$$

For v_R variable in the range 0 to E then clearly the magnitude of the output voltage for a fixed input code $B_1 = B_2 = B_3 = 1$ is variable in the range $-\frac{7}{8}E \leqslant v_o \leqslant 0$ and the 3-bit converter is acting as a single quadrant multiplier (Figure 10.7(b)). Each of the straight lines on this graph represents the variation of the normalized output voltage v_o/V_R for any one of eight fixed input codes and a varying positive reference voltage.

If the reference voltage is allowed to vary in both positive and negative directions the circuit acts as a two-quadrant multiplier. Finally, if provision is made for a fourth bit to be attached to the input code which functions as a sign bit, then a four-quadrant multipler is obtained.

10.6 Current switching DACs

The previously described converters are inherently slow due to the settling time of the operational amplifier. An improvement in operating speed can be achieved by the use of the current switching circuit illustrated in Figure 10.8. The increase in speed is due to a transfer of current rather than turning FETs on and off.

The transistor T_1 is normally biased into saturation, and in the absence of the diode the current I passing through the transistor to the current summing point at the input to the operational amplifier is $I \approx V/2R$.

If logical 1 $(+5\,V)$ is applied to the diode D_1 then it is reverse biased and the current I passes through the transistor. If, on the other hand, the diode input lead is grounded (logical 0) then it is forward biased and, providing the forward resistance of the diode $R_f \lll R_t$, the resistance between collector and emitter of the transistor, the current I is diverted to ground.

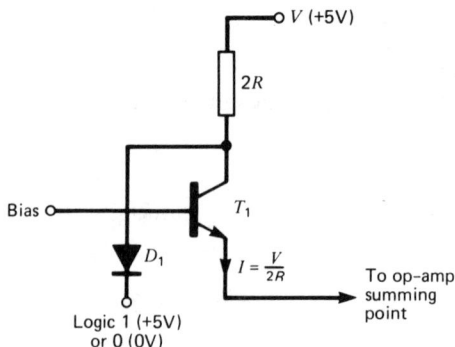

Figure 10.8 Current switching arrangements for a DAC

10.7 Converter coding

Digital data input to a DAC can be represented in a number of different forms, the simplest being a pure binary representation. In converter applications it is convenient to employ a fractional binary representation where each input code combination is represented as a fraction of the reference voltage V_R, which in this context is referred to as the full-scale voltage (FS).

A general expression for a fractional binary number is

$$(N_f)_{10} = a_{-1}2^{-1} + a_{-2}2^{-2} + \ldots a_{-n}2^{-n}$$

where $a_{-1}, a_{-2} \ldots a_{-n}$ are the binary coefficients associated with each term.

The weighting of the most significant bit (MSB) in this series is 2^{-1}, the next most significant bit has a weighting of 2^{-2} and the nth bit has a weighting of 2^{-n}.

When all the binary digits in an n-bit word are 1, the decimal representation of the number is

$$(N_f)_{10} = 1 - 2^{-n}$$

and for $n = 4$

$$\begin{aligned}(N_f)_{10} &= 1 - 2^{-4}\\ &= {}^{15}\!/_{16}\end{aligned}$$

In normal usage the input digital code is not expressed in fractional form but the fractional nature of the number is inferred. For example, the interpretation of the 8-bit input code 10010111 is

$$\begin{aligned}(N_f)_{10} &= 1 \times {}^{1}\!/_{2} + 1 \times {}^{1}\!/_{16} + 1 \times {}^{1}\!/_{64} + 1 \times {}^{1}\!/_{128} + 1 \times {}^{1}\!/_{256}\\ &= {}^{151}\!/_{256} = 0.59\end{aligned}$$

Although pure binary is the most frequently used representation in conversion systems, there are a number of other codes which are used for a variety of reasons. For example, BCD with an 8-4-2-1 weighting could be employed when an analogue-to-digital converter (ADC) is used in conjunction with seven-segment displays to provide a decimal readout.

For an ADC with an 8-bit output the word is divided into two 4-bit quads. The least significant bit in the most significant quad (MSQ) has a weighting of 0.1, while the least significant bit of the second and least significant quad has a weighting of 0.01.

It is not uncommon to use a Gray code with an ADC. This type of code has the advantage that only one digit of the code changes when a transition is made from one code combination to the next one. As an example, a shaft encoder is used to translate an angular displacement into a digital code, as illustrated in the table given in Figure 10.9(*a*). This table illustrates that when the angular displacement is 157½° the output binary code becomes 0111, and when the displacement has increased to 180° the output binary code makes a transition to 1000. At this transition all the encoded digits should change simultaneously. If, however, it happens that due to an overlap on the encoding disc, bit S changes before bits P,Q and R, an intermediate output of PQRS = 0110 occurs, corresponding to an angular displacement of 135°. This is clearly a spurious output. Disc overlaps of this kind can be avoided by using a Gray code disc in which only one digit changes at any given angular displacement.

A 16-combination Gray code can be developed from the K-map shown in Figure 10.9(*b*). Adjacent cells on the K-map represent Boolean terms that differ in one digit place only. If a continuous path is traced through the K-map, as illustrated on the diagram, each cell combination of four binary digits differs from the preceding one in one digit place only, and the Gray code generated is shown in the right-hand column of the table shown in Figure 10.9(*a*).

Angular displacement (degrees)	Binary coded decimal				Gray code				Angular displacement (degrees)	Binary coded decimal				Gray code			
	P	Q	R	S	A	B	C	D		P	Q	R	S	A	B	C	D
0	0	0	0	0	0	0	0	0	180	1	0	0	0	1	1	0	0
22½	0	0	0	1	0	0	0	1	202½	1	0	0	1	1	1	0	1
45	0	0	1	0	0	0	1	1	225	1	0	1	0	1	1	1	1
67½	0	0	1	1	0	0	1	0	247½	1	0	1	1	1	1	1	0
90	0	1	0	0	0	1	1	0	270	1	1	0	0	1	0	1	0
112½	0	1	0	1	0	1	1	1	292½	1	1	0	1	1	0	1	1
135	0	1	1	0	0	1	0	1	315	1	1	1	0	1	0	0	1
157½	0	1	1	1	0	1	0	0	337½	1	1	1	1	1	0	0	0

(*a*)

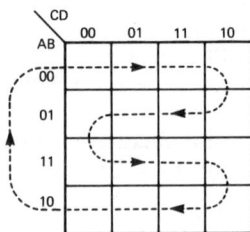

(*b*)

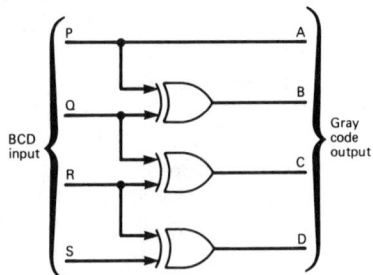

(*c*)

Figure 10.9 (*a*) Digital encoding for angular displacement using BCD and Gray code (*b*) Generation of Gray code using K-map (*c*) Combinational code converter, BCD to Gray code

If necessary, a simple combinational code converter can be employed to translate from binary to Gray code. The Boolean equations for the conversion are

$$A = P \qquad C = Q \oplus R$$
$$B = P \oplus Q \qquad D = R \oplus S$$

and the implementation of these equations appears in Figure 10.9(c).

10.8 Analogue polarity

The DACs described earlier in this chapter are *unipolar* in that they provide a single polarity voltage at the output, and for that reason are referred to as *unipolar DACs*. In a similar way, an ADC that can only handle analogue signals of one polarity is called a *unipolar ADC*. There are, however, *bipolar DACs* available on the market which can convert signed binary numbers into voltages of either polarity. For conversion of bipolar analogue signals into a digital code, the MSB of the generated code is used to indicate the polarity of the input voltage. In the case of a negative voltage the MSB of the digital code is a 1, and for a positive voltage it is 0.

The codes most frequently used for bipolar conversion are sign and magnitude, 2's complement and offset binary. The table given in Figure 10.10 lists each of these codes for four binary digits.

Number	Sign and magnitude	2's complement	Offset binary
+7	0,111	0,111	1,111
+6	0,110	0,110	1,110
+5	0,101	0,101	1,101
+4	0,100	0,100	1,100
+3	0,011	0,011	1,011
+2	0,010	0,010	1,010
+1	0,001	0,001	1,001
0	0,000	0,000	1,000
0	1,000		
−1	1,001	1,111	0,111
−2	1,010	1,110	0,110
−3	1,011	1,101	0,101
−4	1,100	1,100	0,100
−5	1,101	1,011	0,011
−6	1,110	1,010	0,010
−7	1,111	1,001	0,001
−8		1,000	0,000

Figure 10.10 Digital codes used in converters activated by bipolar signals

Sign and magnitude representation is conceptually the simplest way of expressing bipolar analogue quantities digitally, but it has the disadvantage that it is not easy to use computationally, and it is harder to interface with digitally because it requires either additional hardware or software. The 2's complement representation is extensively employed. It is easy to handle computationally since the process of subtraction can be performed as addition in a digital machine. Unfortunately there is a major bit transition occurring between 0 and -1 where all the digits in the code have to change simultaneously. This can lead to the generation of unwanted spikes. Offset binary is perhaps the easiest code to use in a conversion system. Essentially it is pure binary with its zero offset to negative full scale. It has an unambiguous code for zero and it can be derived from the 2's complement code by inverting the most significant bit.

10.9 Resolution

A DAC generates a discrete analogue signal defined by an input digital code. For a 4-bit input code there are 16 corresponding analogue voltages that can be generated within the range of the reference voltage 0 to V_R. Since the converter only produces discrete voltages there is a limit to the resolution with which the digital code can generate a specified analogue signal.

Resolution is defined as the smallest change in the analogue output voltage that can occur as a consequence of the addition of 1 to the LSB of the binary input code. It is always equal to the weight of the LSB of the input codeword; for example, with a 4-bit input word the weight of the LSB is 2^{-4} and the resolution

$$R = \frac{1}{2^4} = .0625$$

which also represents the step size of the converter. When 0001 is added to the input binary word the analogue voltage increases by .0625 V.

More usefully, resolution can be expressed as a percentage of the nominal full-scale output voltage. With a 10 V 4-bit DAC the actual full-scale output voltage is 9.94 V and the percentage resolution is given by

$$\%R = \frac{\text{Step size}}{\text{Nominal FS o/p volts}} \times 100$$

$$= 6.25\%$$

As the number of bits in the input word increases, the percentage resolution decreases and the nominal full-scale output voltage is approached by the actual full-scale output voltage. For a 10 V, 8-bit converter the actual full-scale output voltage = $(255/256) \times 10 = 9.96$ V and the percentage resolution is

$$\%R = \frac{.04}{10} \times 100$$

$$= 0.4\%$$

Clearly, in this example, either the nominal or actual full-scale output voltage can be used to calculate the percentage resolution.

The resolution of a BCD converter is larger, and consequently worse, than the resolution of a binary converter having the same number of input bits. For the 8-bit BCD converter there are 100 steps and the percentage resolution is

$$\%R = \frac{1}{100} \times 100$$

$$= 1\%$$

It should be observed that resolution is a function of the number of bits in the input code and does not depend upon the excellence of the design of the converter. Resolution can only be improved by increasing the number of binary digits in the input word.

10.10 Errors in D/A converters

The actual performance of a DAC differs from its predicted performance because of the occurrence of three different types of errors. These are offset, gain, and linearity errors.

The offset error is the difference between the voltage output of the DAC when the input digits are all zero and the voltage that should have been obtained in this condition. The error produces a vertical shift of the bar graph, as illustrated in Figure 10.11, which may be either positive or negative. Reduction of the offset error to zero can be achieved by applying all the zeros at the converter input and then adjusting the analogue output voltage to zero with a trimming potentiometer. This can be used to trim out the error at a particular temperature, but, since offset error is temperature dependent, a change in temperature will cause a recurrence of the error.

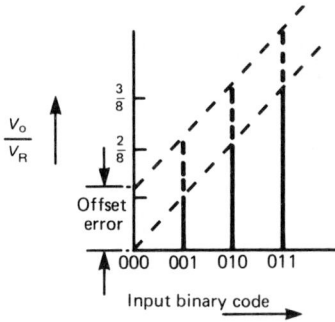

Figure 10.11 Offset error

Gain error results in a rotation of the converter characteristic in either the positive or negative direction. The value of the error varies and depends upon the input codeword. Gain error is illustrated in Figure 10.12.

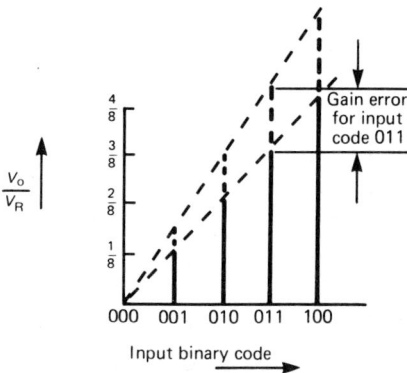

Figure 10.12 Gain error

To eliminate gain error all the 1's are applied at the digital input and the analogue output voltage is trimmed to $(FS - 1LSB)$. Since the error is temperature dependent the adjustment has to be repeated if the temperature changes.

The *non-linearity* of a converter characteristic can be defined in terms of the maximum deviation from a straight line passing through the two end points of the characteristic. An example of linearity error is illustrated in Figure 10.13.

Differential linearity error is defined as the deviation from the analogue difference between any two adjacent code values on the converter characteristic. It will be recalled that, ideally, the analogue difference is an amount equivalent to the LSB $(FS/2^n)$.

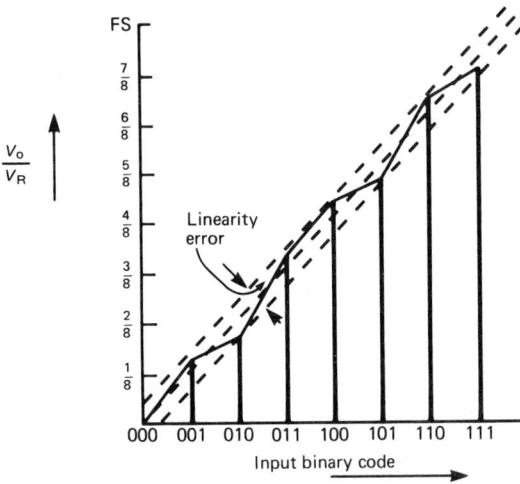

Figure 10.13 Linearity error

For a specified differential linearity error of $\pm$ LSB/2 the discrete voltage difference between two adjacent codes is never less than $\frac{1}{2}$ (FS/2^n) or greater than $\frac{3}{2}$ (FS/2^n).

Providing there is no change in the slope of the transfer characteristic due to non-linearity, it is said to be *monotonic*. In this case the analogue output of the DAC increases, or, at worst, remains the same as the input digital code is incremented. If the differential linearity error is greater than $\pm$ (FS/2^n) it is possible for the analogue output voltage to decrease when the input digital code is incremented, although this will not necessarily always be the case. When a change in the sign of the slope of the transfer characteristic occurs, it is said to be *non-monotonic*. To guarantee monotonicity, the differential linearity error must be less than $\pm$ (FS/2^n).

10.11 Settling time

When a change of code occurs at the input of a DAC a finite time will elapse before the output voltage falls within $\pm$ $\frac{1}{2}$LSB of the final predicted value. This time is referred to as the *settling time*. The concept of settling time is illustrated in Figure 10.14, where the initial input code to a 3-bit DAC is 000 and a transition is made to the code 110. The final predicted value of the output voltage is $3V_R/4$ and the settling time is defined as the time taken for the final value to fall within the analogue voltage range of $(\frac{3}{4} \pm \frac{1}{16})$ V_R.

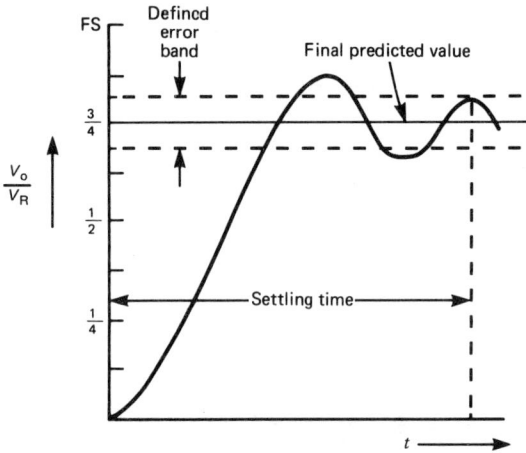

Figure 10.14 Output response of a DAC to a change of input code, showing the concept of settling time

The dynamic behaviour of a DAC is also affected when individual analogue switches within the converter have different switching times. For example, in a 3-bit converter a change of input code from 100 to 011 requires a simultaneous change of all three analogue switches. If the most significant bit of the input code is switched first, then temporarily the input code will be 000 and the output voltage of the converter will be driven towards zero voltage, thus generating a negative spike of voltage, usually referred to as a *glitch*.

10.12 Interfacing a DAC to a source of digital data

In practical situations a DAC has to be interfaced to the device supplying the digital data. The function of the DAC is then to hold a particular analogue voltage at the output until it is commanded to do otherwise by a change in the input word. This requirement can be achieved by using the arrangement shown in Figure 10.15.

Digital data is fed to the data inputs of an 8-bit latch which consists of an array of D-type flip-flops. When an enable signal is received, the digital data is transferred to the outputs of the D-type flip-flops and hence to the data inputs of the DAC. The analogue voltage appearing at the output of the DAC reaches its steady state value after a time period equal to the settling time. Clearly, there is a limit to the rate at which digital data can be converted, and this is governed by the maximum settling time of the device.

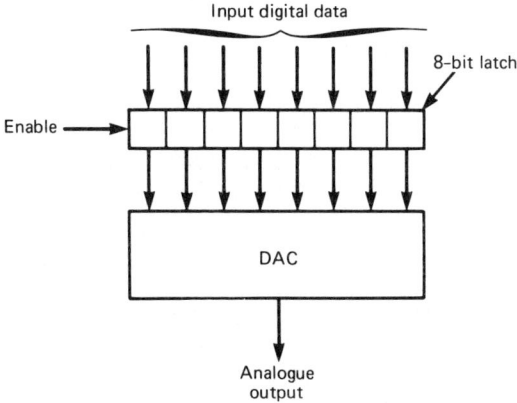

Figure 10.15 The input of parallel data

If the digital data is supplied in serial form it can be clocked into a serial in, parallel out (SIPO) shift register whose outputs are connected to the inputs of an 8-bit latch. Data can now be released from this latch in the manner previously described.

When the digital data to be converted is output from an 8-bit microprocessor and the DAC is, for example, a 10-bit device, the data is provided by two consecutive output words from the processor. The first 8-bit word supplied provides the eight least significant bits for the converter and the two most significant bits for the DAC are provided by the two least significant bits of the second 8-bit word output from the processor.

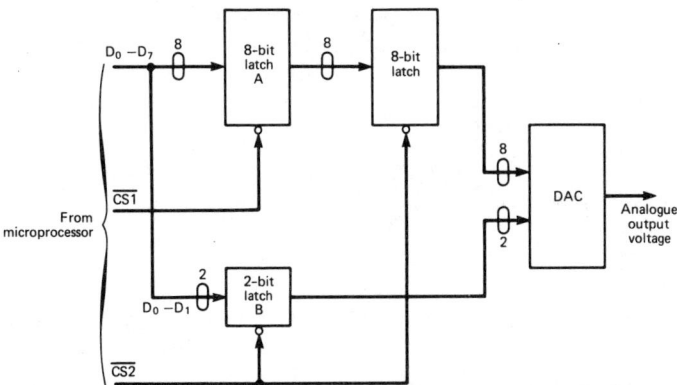

Figure 10.16 Supplying a 10-bit input to a DAC from a microprocessor using a system of double buffering

A system of *double buffering* is then employed in order to ensure that all ten input bits arrive at the DAC simultaneously. The method is illustrated in Figure 10.16. The first eight bits are transferred from latch A to latch B by the chip select signal $\overline{CS1}$ and the 10-bit word is released from latches B and C by the arrival of the second chip select signal $\overline{CS2}$, thus ensuring the simultaneous arrival of all ten bits at the input of the DAC.

10.13 The principles of analogue-to-digital conversion

An ADC performs two basic operations, quantization and coding. Quantization can be described as the process of transforming a continuous signal into one of several equally spaced discrete values, while coding can be regarded as the allocation of a binary code to each of the discrete values into which the analogue signal can be transformed.

A combination of quantization and coding produces the non-linear characteristic of an ADC shown in Figure 10.17. The quantizer staircase function shown in Figure 10.17 is the best approximation that can be made to a straight line passing through the origin and the full-scale point. If the number of divisions into which the analogue voltage is divided is increased, then the

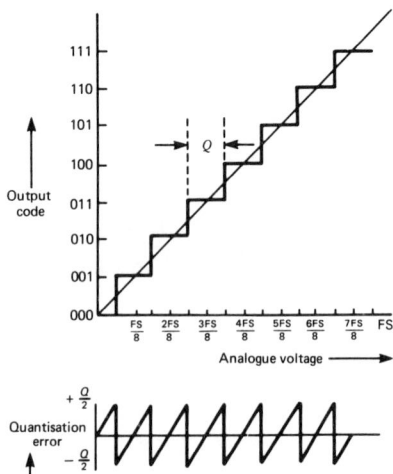

Figure 10.17 Characteristic of an ADC

staircase characteristic will be a closer approximation to this straight line.

The converter produces the same output codeword for a given range of analogue input voltage Q. This range is termed the *analogue quantization size*, or *quantum*. If the full-scale analogue voltage $V_a = 10\,\text{V}$ is to be transformed into a 3-bit binary code then the value of $Q = 10/8 = 1.25\,\text{V}$. In general, Q is given by the equation

$$Q = \frac{\text{FS}}{2^n}$$

For a given full-scale analogue voltage FS, the variation of quanta size Q with the number of bits n in the output code is illustrated in the table shown in Figure 10.18.

No. of output bits n	FS V	Q V
2	10	2.50
3		1.25
4		0.63
5		0.31
6		0.16
7		0.08
8		0.04
9		0.02
10		0.01

Figure 10.18 Variation of quanta Q with number of bits in the output code

If the converter input is moved through the entire analogue voltage range and the difference is taken between a straight line passing through the origin and the full-scale point and the staircase characteristic, a sawtooth variation of the quantization error is obtained (see Figure 10.17). This *quantization error* can only be reduced by increasing the number of output codes of the converter for a given full-scale analogue voltage. The output of the ADC can be thought of as the analogue input with *quantization noise* superimposed on it. The noise has a peak value of Q, an average value of zero, and an rms value of $Q/2\sqrt{3}$.

Changes in the digital output of the converter from one code to the next higher (or lower) code occur at *transition points* or decision levels, and these are fixed at levels $\pm\,Q/2$ on either side of the mid-range values of each quantum (FS/8, FS/4, . . . 7FS/8). Because of the quantization process there is an in-built uncertainty about the output code that lies in the range $-\,Q/2 \leqslant 0 \leqslant +\,Q/2$.

10.14 Sample and hold circuits

Analogue signals which are to be converted into digital form are, in general, varying continuously with time. If the conversion time of the ADC is long, and the rate of change of the analogue signal with respect to time is fast, significant changes in the magnitude of the analogue signal will occur during the conversion process. To eliminate the effect of such changes, the ADC can be preceded by a *sample and hold* circuit (see Figure 10.19(*a*)) whose function is to hold the input analogue voltage to the converter constant during the conversion process.

(*a*)

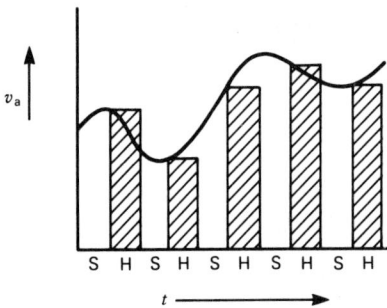

(*b*)

Figure 10.19 (*a*) An ADC system incorporating a sample and hold circuit (*b*) The sampling process

After sampling and conversion, the instantaneous values of the analogue signal at the hold time are digitally represented at the output of the converter by a series of *n*-bit words equally spaced in time. The digital output of the converter can then be fed to a digital machine such as a microprocessor, where it is processed and, if required, the processed data can be reconverted into analogue form by a DAC.

An important requirement in sampled data systems is that the signal should be sampled at a rate which satisfies the *sampling theorem*. This states:

If a continuous, bandwidth limited signal contains no frequency components higher than f_a, the original signal can be recovered without loss of information if it is sampled at a rate of at least $2f_a$ samples/sec.

Failure to do so leads to the phenomenon known as *aliasing*.

A sample and hold circuit, as its name suggests, has two modes of operation. They are (a) sample, and (b) hold. In the sample mode the output of the circuit $V_a = v_a$ where v_a is the instantaneous value of the input signal. On receipt of the hold command, the output of the circuit is ideally held constant at a value equal to the input signal at the instant of receiving the hold command, for the period of the hold time. The behaviour of the circuit is illustrated in Figure 10.19(*b*), where the time scale is divided into sample (S) and hold (H) periods. In the sample periods the output voltage tracks, or follows, the input signal. At the commencement of the hold period the output voltage is held at a constant value equal to the value of the input voltage at the start of the period. The voltage output during the hold periods is represented by the hatched portions of the graph. It is clear that the length of the hold period must be at least equal to the conversion time of the ADC.

The simplest form of sample and hold circuit consists of a switch S and a capacitor C, connected as shown in Figure 10.20(*a*). When the switch is closed the output voltage V_a follows the input voltage v_a. On opening the switch, the voltage across the capacitor is, ideally, maintained at the value of the input voltage just prior to the instant of switching. In practice, the capacitor is connected across the input resistance of the next stage and will discharge through this resistance during the hold period, resulting in a decay of voltage across the capacitor. The çircuit can be isolated from the next stage by a unity gain buffer whose input resistance is extremely large (Figure 10.20(*b*)) and, as a consequence, the capacitor voltage remains sensibly constant during the hold period.

When the switch is closed, the large capacitance C is connected directly across the source. It may load this stage up and generate parasitic oscillations, so it is common practice to isolate the sample and hold circuit from the previous stage by a unity gain buffer (Figure 10.20(*c*)).

The switch S may be a relay (for slow waveforms), a bipolar transistor or an FET controlled by its gate voltage, as illustrated in Figure 10.20(*d*). On receipt of the hold command, the FET switch is opened, and the sample and hold circuit enters the hold period.

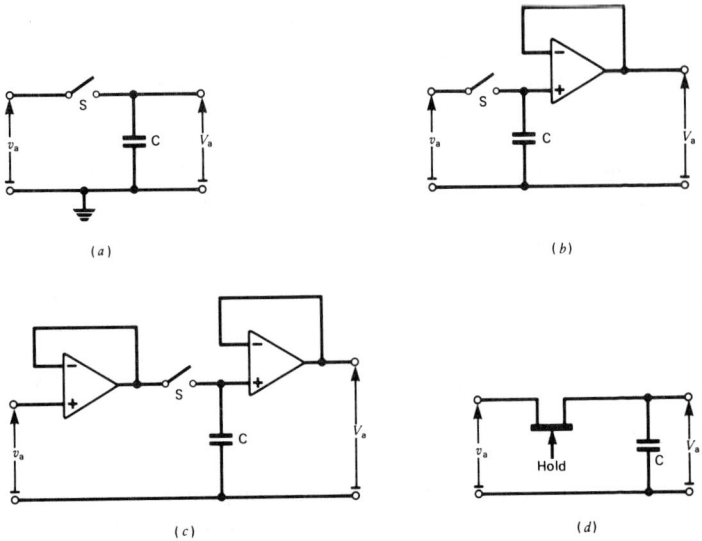

Figure 10.20 (a) Basic sample and hold circuit (b) With isolating unity gain buffer (c) With isolation of both source and output (d) With electronic switch

A practical analogue-to-digital conversion system incorporating a sample and hold circuit is illustrated in Figure 10.21. The low pass filter (LPF) is used to limit the bandwidth of the input analogue signal to meet the requirements of the sampling theorem. Logical control of the system is provided by the digital control circuit, which generates the HOLD command for the FET switch simultaneously with the start of conversion (SOC) signal for the ADC. At the end of conversion, the converter generates the EOC signal which initiates the $\overline{\text{HOLD}}$ = SAMPLE signal in the control circuit and supplies it to the FET switch.

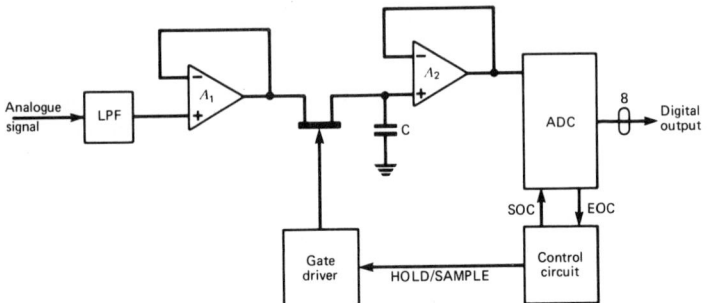

Figure 10.21 Practical ADC system

There are three important factors to be considered in the operation of a sample and hold circuit within an ADC system such as the one illustrated in Figure 10.21. First, there is the *acquisition time*, which is defined as the time taken from the arrival of the sample signal to the point at which the output falls within a specified allowable error band. In essence, it represents the time taken to charge up capacitor C through the low output resistance of the unity gain buffer A_1.

When the FET switch is opened by the HOLD command, the capacitor discharges through the 'open' resistance of the switch, and the input resistance of the other unity gain buffer A_2. Both these resistance values are high, and the decay of capacitor voltage is small. This decay occurring during the ADC conversion time is referred to as the *droop voltage*.

Finally, *aperture time* is defined as the time between the transition of the control signal to the hold state and the opening of the FET switch. It may be regarded as the time taken after the arrival of the hold signal for the output voltage of the sample and hold circuit to become independent of the input voltage.

10.15 Comparators

An important operation carried out during an analogue-to-digital conversion is the comparison of two voltages. The operation is performed by a device called a *comparator*, whose function is to provide a digital output signal which has only two logical values 1 and 0.

A comparator consists of a subtraction circuit whose input voltages are V_1 and V_2 respectively, followed by a high gain amplifier (see Figure 10.22(a)). The subtractor operating in conjunction with the high gain amplifier is, in effect, an operational amplifier, designed to work in the differential mode. Ideally, the combination should have the characteristic shown in Figure 10.22(b) where the output voltage $A\,\Delta V$ makes a direct transition from $-V_S$ to $+V_S$ when $V_1 - V_2$ passes through zero. In practice, the characteristic obtained is shown in Figure 10.22(c) and it will be observed that in this case the transition endpoints are at $\pm V_T$ where V_T is termed the *threshold voltage* and is of the order of 1 mV.

The equations of this characteristic are:

$$V_1 - V_2 \leqslant -V_T \; ; \quad \text{output saturates at } -V_S$$
$$-V_T \leqslant V_1 - V_2 \leqslant +V_T \; ; \quad A\,\Delta V = A\,(V_1 - V_2)$$
$$V_1 - V_2 \geqslant +V_T \; ; \quad \text{output saturates at } +V_S$$

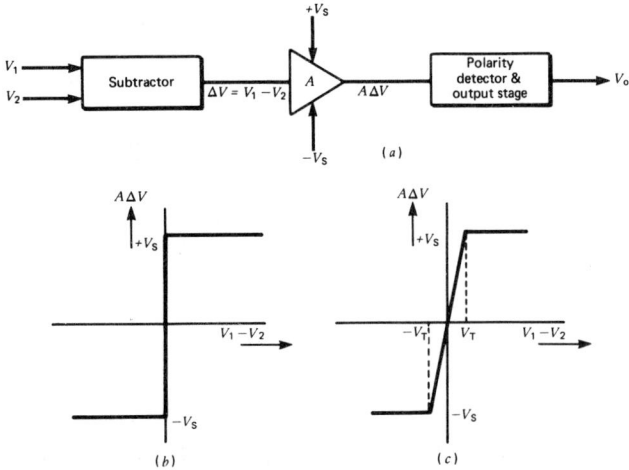

Figure 10.22 (a) Basic block diagram of a comparator (b) Ideal characteristic, and (c) practical characteristic of subtractor, in conjunction with a high gain amplifier

The output levels from the high gain amplifier are not compatible with TTL logic. Conversion to TTL levels is provided by the polarity detector and output stage. The characteristic of the comparator as seen at the output is shown in Figure 10.23.

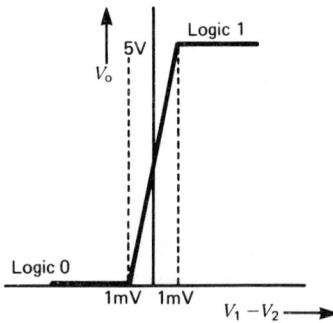

Figure 10.23 Characteristic of TTL compatible comparator

10.16 Counting ADCs

In this method of conversion the analogue voltage V_a is repeatedly compared to the output voltage V_d of a reference DAC. The output of the DAC is controlled by the comparison and continues to increase until it is in correspondence with the input analogue voltage ($V_d = V_a$). At this point the input to the DAC is the binary representation of the input analogue voltage.

The system used when implementing this method of conversion will be explained in conjunction with Figure 10.24(a). On receipt of an SOC signal, the counter is cleared, hence $V_a > V_d$ since $V_d = 0\,V$. The output of the comparator is logical 1 and provides an enabling signal for the counter via the enabling gate g_1. As the counter is clocked, it counts up and its output is the digital input to the reference DAC. This output is converted into an analogue voltage V_d which is gradually increasing in steps with each incoming clock pulse, as illustrated in Figure 10.24(b). When $V_d \geqslant V_a$ the output of the comparator changes to logical 0, disables the counter and provides an end of conversion signal (EOC). The output of the counter at this instant is the required binary representation of V_a.

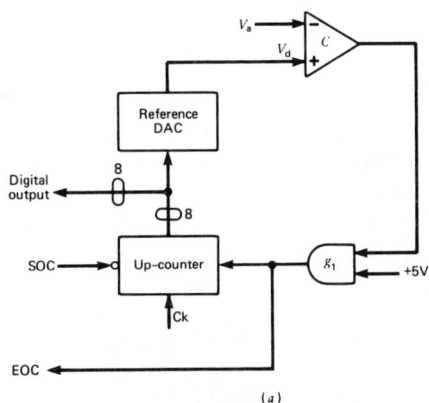

(a)

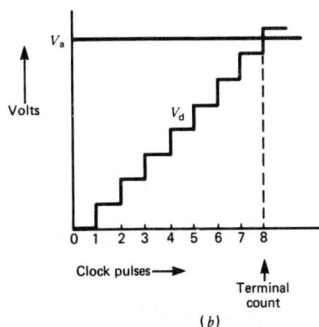

(b)

Figure 10.24 (a) Block diagram of a counting ADC (b) Graphical description of the operation

The major advantage of this converter is that only the DAC and the comparator affect its accuracy. However, the conversion time is variable and depends upon the magnitude of the analogue voltage V_a to be converted. The periodic time of the clock must be at least equal to the sum of the DAC's settling time and the comparator response time.

In a microprocessor system the SOC signal would be generated by the processor and would be transferred to the ADC via an output port. At the end of conversion the EOC signal generated by the ADC is returned to the processor. This signal initiates an enabling signal for the input port temporarily storing the digital data, which is then transferred to the accumulator.

A limitation of the counting ADC described above is that at the start of every conversion the counter has to be cleared to zero even if the change of analogue input signal between consecutive conversions is small. By replacing the up-counter with an up-down counter the ADC can be converted into a tracking ADC and the disadvantage previously identified is overcome.

The output of the comparator is now used as the up/down control of the converter. When $V_a > V_d$ the counter is being incremented, and when $V_a < V_d$ it is being decremented. The error signal is being continuously monitored by the comparator and consequently the output signal of the DAC follows the analogue input signal. If the analogue signal is changing only slowly with time, so that in one clock cycle the change in V_a is small, then the time required to produce a new binary representation will be small. On the other hand, if large changes are taking place in V_a then the conversion time is not significantly different from that of the previously described converter.

10.17 Principle of successive approximation conversion

Figure 10.25(a) shows a 4-bit DAC operating with a reference voltage V_R and having an output voltage V_d. The maximum value of the output voltage occurs when the input binary code is 1111 and then $V_{d\,max} = 15V_R/16$. Comparison of the output of the DAC is to be made with the analogue voltage to be converted, $V_a = 25V_R/32$.

Step 1 Input binary code = 1000 and generated $V_d = 8V_R/16$, $V_a > 8V_R/16$. Hence leave MSB unchanged and make second most significant digit of input binary code = 1.

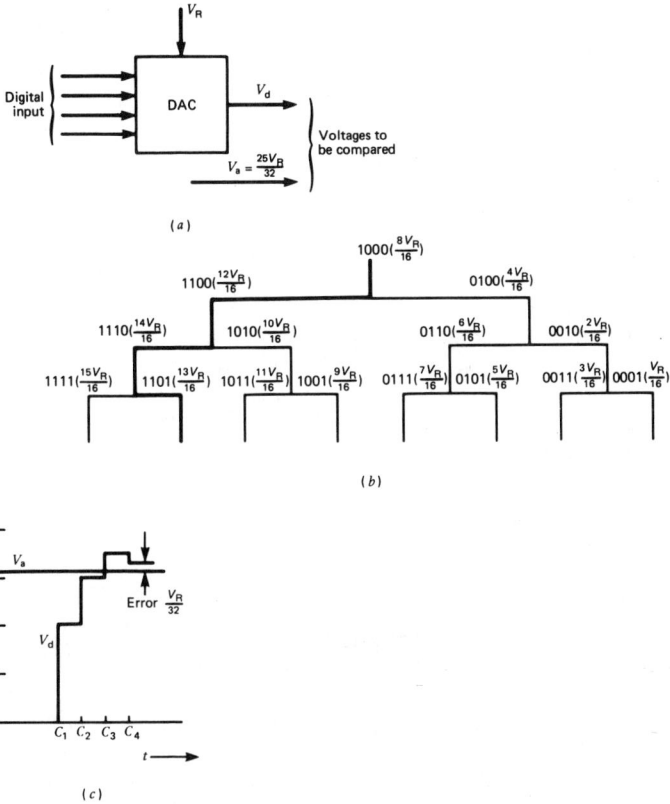

Figure 10.25 Successive approximation technique (a) Block diagram (b) Tree structure (c) Graphical description

Step 2 Input binary code = 1100 and generated $V_d = 12V_R/16$, $V_a > 12V_R/16$. Hence leave second most significant digit unchanged and make third most significant digit of input binary code = 1.

Step 3 Input binary code = 1110 and generated $V_d = 14V_R/16$, $V_a < 14V_R/16$. Hence change third most significant digit back to zero and make the LSB = 1.

Step 4 Input binary code = 1101 and generated $V_d = 13V_R/16$, $V_a < 13V_R/16$. This is the final comparison possible with four digits.

With only four input digits available for conversion by the DAC the closest approximation has been made to the incoming analogue voltage by this technique of successive approximation. The method is illustrated pictorially by the tree structure shown in Figure 10.25(b) and the path taken through this structure for the example given above is emphasized by a heavy black line. A plot of both V_d and V_a against time is shown in Figure 10.25(c) and it will be observed that the value of the error obtained is $V_R/32$ and that the number of comparisons made is equal to the number of binary digits in the input word.

There are two types of ADC employing the successive approximation technique that are in common use. The first one, termed a *post-subtractive* converter behaves in the way described in Figure 10.25(c). If the difference $V_a - V_d$ is negative after comparison, the last fraction of the reference voltage added is retained and the next lower fraction of V_R is subtracted. The alternative technique employed is referred to as *pre-subtractive* conversion and is illustrated in Figure 10.26. In this case, if the result of the comparison is negative the last fraction of the reference voltage to be added has to be subtracted before the next lower fraction can be added.

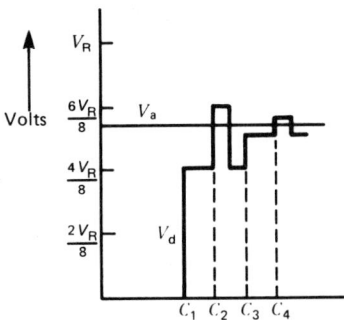

Figure 10.26 Pre-subtractive conversion

10.18 Successive approximation ADC

Figure 10.27(a) shows the circuit arrangements for a 4-bit successive approximation ADC. It consists of a seven-state shift register T, U, V, W, X, Y, Z, a 4-bit DAC operating with a reference voltage V_R, four D-type flip-flops P, Q, R, S, which provide the input to the DAC and also to the 4-bit output register, and a comparator C. The gate g_1 and JK flip-flop A provide the control logic for implementing the successive approximation method.

The circuit is initialized by the SOC signal which presets FFT and clears FFs U, V, W, X, Y, and Z. When FFT is preset to 1 there is a negative-going transition at the $\overline{T}$ output which clears

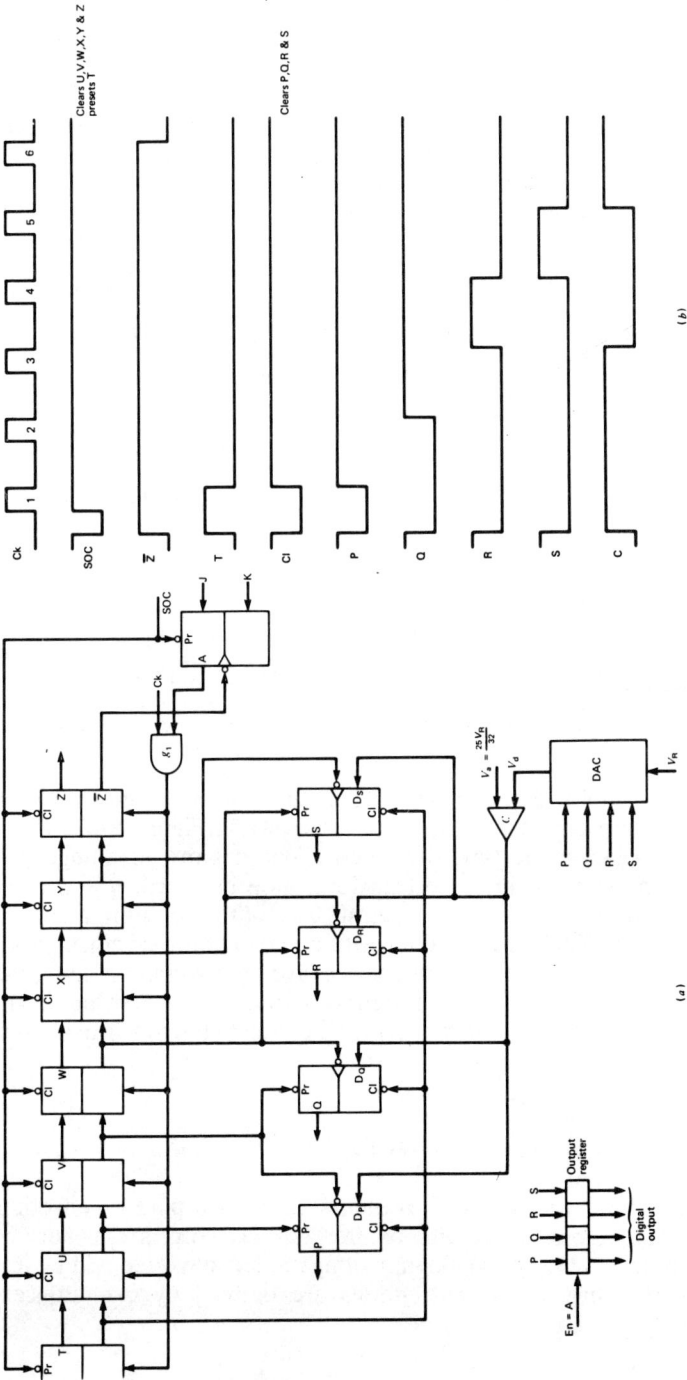

ure 10.27 (a) Successive approximation register (b) Timing diagram for successive approximation ADC

FFs P,Q,R, and S so that the input to the DAC and the output register is PQRS = 0000. The output of the DAC $V_d = 0$, hence $V_a > V_d$ and the output of the comparator C = 1. Additionally, the SOC signal presets FFA, enabling g_1, and allowing the passage of clock pulses to the shift register. Circuit operation is now under control of the clock and transitions are made on the trailing edge of the clock pulses in the sequence shown below. In the analysis it is assumed that $V_a = 25V_R/32$.

Ck 1; U:0→1; Ū: 1→0; P:0→1; DAC input PQRS = 1000;

$V_a = 25V_R/32$; $V_d = 16V_R/32$; $V_a > V_d$; C = 1; $D_P = 1$

Ck 2; V̄: 1→0; P: 1→1; Q:0→1; DAC input PQRS = 1100;

$V_a = 25V_R/32$; $V_d = 24V_R/32$; $V_a > V_d$; C = 1; $D_Q = 1$

Ck 3; W̄: 1→0; Q: 1→1; R: 0→1; DAC input PQRS = 1110;

$V_a = 25V_R/32$; $V_d = 28V_R/32$; $V_a < V_d$; C = 0; $D_R = 0$

Ck 4; X̄: 1→0; R: 1→0; S: 0→1; DAC input PQRS = 1101;

$V_a = 25V_R/32$; $V_d = 26V_R/32$; $V_a < V_d$; C = 0; $D_S = 0$

Ck 5; Ȳ: 1→0; S: 1→0; DAC input PQRS = 1100;

$V_a = 25V_R/32$; $V_d = 24V_R/32$; $V_a > V_d$; C = 1;

Ck 6; Z̄: 1→0; A: 1→0; Output register clocked by A and outputs 1100

This process will be repeated when the next SOC signal arrives. The timing digram for the circuit is shown in Figure 10.27(b).

An ADC of the type described above has the advantage of a short conversion time. To transform an n-bit word, $n + 2$ clock pulses are required, whereas in the case of the counting ADC the conversion time was variable, with a maximum of 2^n clock pulses required. Adding a bit to the successive approximation converter increases the clock pulses required by one to $n + 3$, while in the counting ADC the number of clock pulses required for maximum conversion time has increased to 2^{n+1}.

10.19 Parallel A/D conversion

The fastest type of ADC available employs a parallel encoding technique, where all bits of the output word are generated simultaneously. The analogue voltage to be converted will lie in a voltage range whose extremities are defined by two adjacent discrete voltage levels. Each of these levels represents the threshold voltage for a pair of comparators.

In this type of converter a potential divider is used to split the reference voltage V_R into $(2^n - 1)$ equally spaced, discrete levels, where n is the number of bits in the output word. For $n = 3$ there

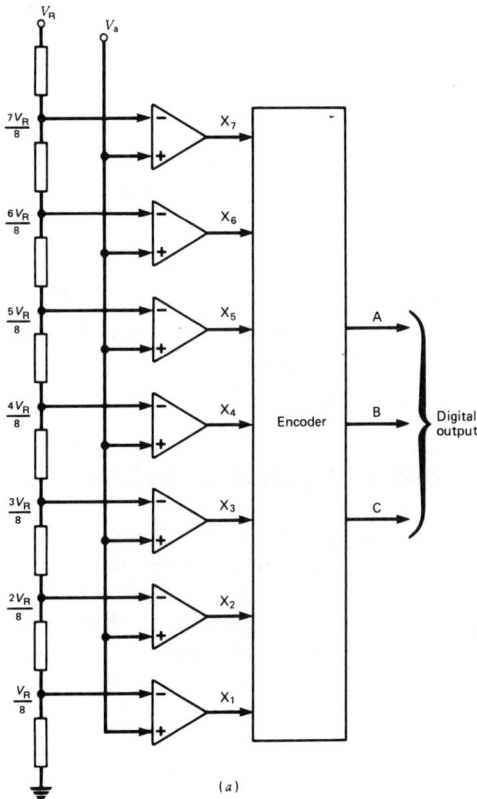

(a)

Encoder inputs							Encoder outputs		
X_7	X_6	X_5	X_4	X_3	X_2	X_1	A	B	C
0	0	0	0	0	0	0	0	0	0
0	0	0	0	0	0	1	0	0	1
0	0	0	0	0	1	1	0	1	0
0	0	0	0	1	1	1	0	1	1
0	0	0	1	1	1	1	1	0	0
0	0	1	1	1	1	1	1	0	1
0	1	1	1	1	1	1	1	1	0
1	1	1	1	1	1	1	1	1	1

(b)

Figure 10.28 (a) Parallel A/D conversion (b) Encoder truth table are seven discrete levels starting at $7V_R/8$ at the top end of the range, the lowest level being $V_R/8$. These discrete voltage levels represent the reference voltage input to a series of comparators (see Figure 10.28(a)) while the other input is the analogue voltage V_a. If V_a is greater than the reference level applied to a comparator its output is 1, otherwise it is 0. The outputs of the comparators X_1, X_2 ... X_7 are applied to the inputs of an encoding circuit, similar to that described in Chapter 2, and whose output is a 3-bit code.

A truth table for the encoding circuit is given in Figure 10.28(b). For the condition $4V_R/8 < V_a < 5V_R/8$, X_1 to X_4 are all equal to 1 and the rest of the outputs X_5 to X_7 are equal to zero. An examination of the truth table reveals that the output code of the ADC is ABC = 100.

As one might expect, with parallel conversion the amount of hardware needed increases rapidly with the length of the output word. If $n = 8$ then a total of 255 comparators is required. The use of this type of converter is limited to those cases where high speed is essential and poor resolution is acceptable.

10.20 Errors in analogue-to-digital converters

Quantization error, dealt with in Section 10.13 of this chapter, is illustrated in Figure 10.29. For a 3-digit converter the output code is 001 for any voltage lying in the range $(FS/2^3 - Q/2)$ to $(FS/2^3 + Q/2)$, where Q = the quantum or voltage difference between two adjacent transition levels. The quantization error E follows a

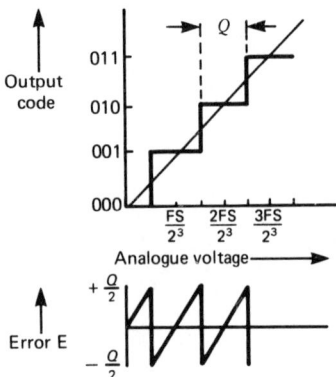

Figure 10.29 Quantization error

sawtooth pattern through the whole of the analogue voltage range and can have a value in the range 0 to $\pm Q/2$. This error is inherent to the conversion technique employed, and depends upon the resolution of the converter. If the number of steps available in a given voltage range is increased, there will be a reduction in the value of Q and the range of the quantization error. All other errors encountered in ADCs are caused by device imperfections and are due to the presence of offset voltages, variable gain and non-linearities.

10.21 Software implementation of ADCs

To implement the counting ADC described in Section 10.16, the basic hardware required is an 8-bit DAC, a comparator and an 8-bit counter. It is possible to reduce the amount of hardware required when the ADC is used in conjunction with a microprocessor, since one of the internal registers used in a counting mode can replace the external 8-bit counter. The

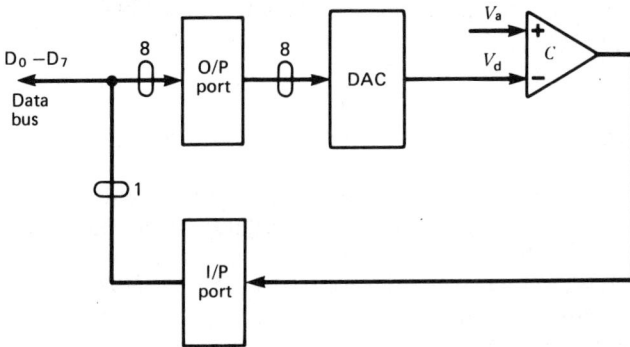

Figure 10.30 Block diagram for a software version of a counting ADC

processor is now dedicated to conversion and cannot be used for any other purpose while conversion is taking place. Although the software approach provides a hardware saving there is a corresponding increase in conversion time since the digital input to the DAC has to be transferred from the microprocessor via an output port, and after comparison the comparator output has to be returned to the processor via an input port. A block diagram illustrating a software conversion scheme is shown in Figure 10.30.

The subroutine CTADC for an 8-bit conversion is listed below:

```
CTADC:  MVI   E,    FFH   ; Set ADC count to −1
        MVI   C,    0FH   ; Set up count in C register
CNT:    INR   E           ; Increment ADC count
        MOV   A,E         ; Move ADC count to acc.
        OUT         01H   ; Output ADC count
LOOP:   DCR   C           ;  ⎫ Delay for comparison
        JNZ         LOOP  ;  ⎬
        IN          02H   ; Check comparator signal
        ANA   A           ; Set flags
        JM          CNT   ; Jump on negative
        RET
```

The ADC count stored in register E is transferred via the accumulator to the output port whose address is 01H and thence to the DAC. After comparison of V_d and V_a the comparator signal is returned to the accumulator via the input port whose address is 02H. The flags are set by the ANA instruction, and if the contents of the accumulator are negative then $V_d < V_a$, and register E is incremented, the process just described then being repeated. A delay is provided by the count in register C while the comparison of V_a and V_d is being made. In this case a purely arbitrary count has been chosen.

Problems

10.1 An 8-bit D/A converter produces an output voltage of 3.164 volts when the input code is 01010001. Determine the maximum value of the output voltage and also the value of the output voltage when the code is 01110010. Also calculate the resolution of the converter and express it in (a) volts, and (b) as a percentage.

10.2 Draw a circuit diagram for an n-bit weighted resistor D/A converter employing an operational amplifier, and show that the expression for the output voltage V_o is given by

$$V_o = (1 + \tfrac{1}{2} + \tfrac{1}{4} + \ldots 1/2^n)\, V_R\, R/R'$$

where V_R = the reference voltage, R is the value of the feedback resistor and R' is the resistance value associated with the most significant digit of the input.

If $V_R = -10\,\text{V}$, $R = 5\,k\Omega$, $R' = 10\,k\Omega$ and $n = 8$ determine the value of the current taken from the reference supply.

Also determine the current entering the summing function and the output voltage for a digital input of 10101110.

10.3 A 12-bit (three-digit) D/A converter has a BCD input code with a full scale of approximately 10 volts. Determine the step size, the percentage resolution and the value of the output voltage V_o for a decimal input of $(721)_{10}$.

Assuming the converter is used with a pure binary input, determine the percentage resolution and compare with your previous answer.

10.4 A counting A/D converter operating at a clock frequency of 1 MHz uses a 12-bit counter. The D/A converter has a full-scale output voltage of 10.24 V and the threshold voltage for the comparator is 1 mV. Draw a block diagram for the converter and determine the digital equivalent for an input analogue voltage of 5.821 V. Additionally, determine the conversion time and resolution of the converter and sketch the output waveform of the D/A converter. Explain how the circuit can be modified in order to reduce the conversion time.

Appendix: Intel 8085A Instruction set summary

(1) Data transfer group

Hex	Mnemonic	Flags
Inter-register transfers		
(r1)←(r2)		
7F	MOV A, A	None
78	MOV A, B	None
79	MOV A, C	None
7A	MOV A, D	None
7B	MOV A, E	None
7C	MOV A, H	None
7D	MOV A, L	None
47	MOV B, A	None
40	MOV B, B	None
41	MOV B, C	None
42	MOV B, D	None
43	MOV B, E	None
44	MOV B, H	None
55	MOV B, L	None
4F	MOV C, A	None ·
48	MOV C, B	None
49	MOV C, C	None
4A	MOV C, D	None
4B	MOV C, E	None
4C	MOV C, H	None
4D	MOV C, L	None
57	MOV D, A	None
50	MOV D, B	None
51	MOV D, C	None
52	MOV D, D	None
53	MOV D, E	None
54	MOV D, H	None
55	MOV D, L	None
5F	MOV E, A	None
58	MOV E, B	None
59	MOV E, C	None
5A	MOV E, D	None
5B	MOV E, E	None
5C	MOV E, H	None
5D	MOV E, L	None

Hex	Mnemonic	Flags
67	MOV H, A	None
60	MOV H, B	None
61	MOV H, C	None
62	MOV H, D	None
63	MOV H, E	None
64	MOV H, H	None
65	MOV H, L	None
6F	MOV L, A	None
68	MOV L, B	None
69	MOV L, C	None
6A	MOV L, D	None
6B	MOV L, E	None
6C	MOV L, H	None
6D	MOV L, L	None
Transfer from memory to register		
(r)←((H)(L))		
7E	MOV A, M	None
46	MOV A, M	None
4E	MOV A, M	None
56	MOV A, M	None
5E	MOV A, M	None
66	MOV A, M	None
6E	MOV A, M	None
Transfer from register to memory		
((H)(L))←(r)		
77	MOV M, A	None
70	MOV M, B	None
71	MOV M, C	None
72	MOV M, D	None
73	MOV M, E	None
74	MOV M, H	None
75	MOV M, L	None

Hex	Mnemonic	Flags
Transfer immediate to register		
(r)←(byte 2)		
3E	MVI A, b2	None
06	MVI B, b2	None
0E	MVI C, b2	None
16	MVI D, b2	None
1E	MVI E, b2	None
26	MVI H, b2	None
2E	MVI L, b2	None
Transfer immediate to memory		
((H)(L))←(byte 2)		
36	MVI, b2	None
Load register pair immediate		
(rh)←(byte 3)		
(rl)←(byte 2)		
01	LXI B, b3 & b2	None
11	LXI D, b3 & b2	None
21	LXI H, b3 & b2	None
31	LXI SP, b3 & b2	None
Register exchange		
(H)←→(D) (L)←→(E)		
EB	XCHG HL←→DE	None

Hex	Mnemonic	Flags	Hex	Mnemonic	Flags

Load accumulator
(A)←((byte 3)(byte 2))

Store accumulator
((byte 3)(byte 2))←A

3A	LDA addr.	None	32	STA. addr.	None

(A)←((rp))

((rp))←A

0A	LDAX B A← M addr. in B,C	None	02	STAX B M←A addr. in B,C	None
1A	LDAX D A←M addr. in D,E	None	12	STAX D M←A addr. in D,E	None

Load registers H and L
(L)←((byte 3)(byte 2))
(H)←((byte 3)(byte 2)+1)

Store registers H and L
((byte 3)(byte 2))←(L)
((byte 3)(byte 2)+1)←(H)

2A	LHLD addr. $\begin{matrix} L \leftarrow M \\ H \leftarrow M+1 \end{matrix}$	None	22	SHLD addr. $\begin{matrix} M \leftarrow L \\ M+1 \leftarrow H \end{matrix}$	None

(2) Arithmetic group

Hex	Mnemonic	Flags
Add register (A)←(A)+(r)		
87	ADD A	All
80	ADD B	All
81	ADD C	All
82	ADD D	All
83	ADD E	All
84	ADD H	All
85	ADD L	All
Add register with carry (A)←(A)+(r)+(CY)		
8F	ADC A	All
88	ADC B	All
89	ADC C	All
8A	ADC D	All
8B	ADC E	All
8C	ADC H	All
8D	ADC L	All
Add immediate (A)←(A)+(byte 2)		
C6	ADI, b2	All
Add immediate with carry (A)←(A)+(byte 2)+(CY)		
CE	ACI, b2	All
Add memory (A)←(A)+((H)(L))		
86	ADD M	All
Add memory with carry (A)←(A)+((H)(L))+(CY)		
8E	ADC M	All

Hex	Mnemonic	Flags
Subtract register (A)←(A)−(r)		
97	SUB A	All
90	SUB B	All
91	SUB C	All
92	SUB D	All
93	SUB E	All
94	SUB H	All
95	SUB L	All
Subtract register with borrow (A)←(A)−(r)−(CY)		
9F	SBB A	All
98	SBB B	All
99	SBB C	All
9A	SBB D	All
9B	SBB E	All
9C	SBB H	All
9D	SBB L	All
Subtract immediate (A)←(A)−(byte 2)		
D6	SUI, b2	All
Subtract immediate with borrow (A)←(A)−(byte 2)−(CY)		
DE	SBI, b2	All
Subtract memory (A)←(A)−((H)(L))		
96	SUB M	All
Subtract memory with borrow (A)←(A)−((H)(L))−(CY)		
9E	SBB M	All

Hex	Mnemonic	Flags
Increment register (r)←(r)+1		
3C	INR A	Z,S,P,AC
04	INR B	Z,S,P,AC
0C	INR C	Z,S,P,AC
14	INR D	Z,S,P,AC
1C	INR E	Z,S,P,AC
24	INR H	Z,S,P,AC
2C	INR L	Z,S,P,AC
Increment memory ((H)(L))←((H)(L))+1		
34	INR M	Z,S,P,AC
Decrement register (r)←(r)−1		
3D	DCR A	Z,S,P,AC
05	DCR B	Z,S,P,AC
0D	DCR C	Z,S,P,AC
15	DCR D	Z,S,P,AC
1D	DCR E	Z,S,P,AC
25	DCR H	Z,S,P,AC
2D	DCR L	Z,S,P,AC
Decrement memory ((H)(L))←((H)(L))−1		
35	DCR M	Z,S,P,AC
Increment register pair (rh)(rl)←(rh)(rl)+1		
03	INX B	None
13	INX D	None
23	INX H	None
33	INX SP	None
Decrement register pair (rh)(rl)←(rh)(rl)−1		
0B	DCX B	None
1B	DCX D	None
2B	DCX H	None
3B	DCX SP	None
Add register pair to HL (H)(L)←(H)(L)+(rh)(rl)		
09	DAD B	CY
19	DAD D	CY
29	DAD H	CY
39	DAD SP	CY
Decimal adjust accumulator		
27	DAA	All

(3) Logic group

Hex	Mnemonic	Flags
AND register $(A) \leftarrow (A) \wedge (r)$		
A7	ANA A	All
A0	ANA B	All
A1	ANA C	All
A2	ANA D	All
A3	ANA E	All
A4	ANA H	All
A5	ANA L	All
AND memory $(A) \leftarrow (A) \wedge ((H)(L))$		
A6	ANA M	All
AND immediate $(A) \leftarrow (A) \wedge (\text{byte 2})$		
E6	ANI, b2	All
Compare register $(A) - (r)$		
BF	CMP A	All
BE	CMP B	All
B9	CMP C	All
BA	CMP D	All
BB	CMP E	All
BC	CMP H	All
BD	CMP L	All
Compare memory $(A) - ((H)(L))$		
BE	CMP M	All
Compare immediate $(A) - (\text{byte 2})$		
FE	CPI b2	All

Hex	Mnemonic	Flags
Exclusive OR register $(A) \leftarrow (A) \oplus (r)$		
AF	XRA A	All
A8	XRA B	All
A9	XRA C	All
AA	XRA D	All
AB	XRA E	All
AC	XRA H	All
AD	XRA L	All
Exclusive OR memory $(A) \leftarrow (A) \oplus ((H)(L))$		
AE	XRA M	All
Exclusive OR immediate $(A) \leftarrow (A) \oplus (\text{byte 2})$		
EE	XRI b2	All
Rotate left $(A_{n+1}) \leftarrow (A_n)$; $(A_0) \leftarrow (A_7)$; $(CY) \leftarrow (A_7)$		
07	RLC	CY
Rotate right $(A_n) \leftarrow (A_{n+1})$; $(A_7) \leftarrow (A_0)$; $(CY) \leftarrow (A_0)$		
0F	RRC	CY
Rotate left through carry $(A_{n+1}) \leftarrow (A_n)$; $(CY) \leftarrow (A_7)$; $(A_0) \leftarrow (CY)$		
17	RAL	CY
Rotate right through carry $(A_n) \leftarrow (A_{n+1})$; $(CY) \leftarrow (A_0)$; $(A_7) \leftarrow (CY)$		
1F	RAR	CY

Hex	Mnemonic	Flags
OR register $(A) \leftarrow (A) \vee (r)$		
B7	ORA A	All
B0	ORA B	All
B1	ORA C	All
B2	ORA D	All
B3	ORA E	All
B4	ORA H	All
B5	ORA L	All
OR memory $(A) \leftarrow (A) + ((H)(L))$		
B6	ORA M	All
OR immediate $(A) \leftarrow (A) + (\text{byte 2})$		
F6	ORI b2	All
Complement accumulator $(A) \leftarrow (\bar{A})$		
2F	CMA	None
Complement carry $(CY) \leftarrow (\overline{CY})$		
3F	CMC	CY
Set carry $(CY) \leftarrow 1$		
37	STC	CY

(4) Branch group

Hex	Mnemonic	Flags

Jump
(PC)←(byte 3)(byte 2)

| C3 | JMP addr. | None |

Conditional jump
(PC)←(byte 3)(byte 2)
if specified condition is true

C2	JNZ addr.	None
CA	JZ addr.	None
D2	JNC addr.	None
DA	JC addr.	None
E2	JPO addr.	None
EA	JPE addr.	None
F2	JP addr.	None
FA	JM addr.	None

Restart and save program counter on stack
(PC)←8*(NNN)

C7	RST 0	None
CF	RST 1	None
D7	RST 2	None
DF	RST 3	None
E7	RST 4	None
EF	RST 5	None
F7	RST 6	None
FF	RST 7	None

Hex	Mnemonic	Flags

Call
Stack←(PC); (SP)←(SP)−2;
(PC)←(byte 3)(byte 2)

| CD | CALL addr. | None |

Conditional call
Stack←(PC);
(PC)←(byte 3)(byte 2);
if specified condition is true

C4	CNZ addr.	None
CC	CZ addr.	None
D4	CNC addr.	None
DC	CC addr.	None
E4	CPO addr.	None
EC	CPE addr.	None
F4	CP addr.	None
FC	CM addr.	None

Move H and L to PC
(PCH)←H; (PCL)←L

| E9 | PCHL | None |

Hex	Mnemonic	Flags

Return
(PCL)←((SP));
(PCH)←((SP)+1);
(SP)←(SP)+2

| C9 | RET | None |

Conditional return
(PCL)←((SP));
(PCH)←((SP)+1);
(SP)←(SP)+2;
if specified condition is true

C0	RNZ	None
C8	RZ	None
D0	RNC	None
D8	RC	None
E0	RPO	None
E8	RPE	None
F0	RP	None
F8	RM	None

(5) Stack, I/O and machine control group

Hex	Mnemonic	Flags
Push register pair ((SP)−1)←(rh); ((SP)−2)←(rl); (SP)←(SP)−2		
C5	PUSH B	None
D5	PUSH D	None
E5	PUSH H	None
Push PSW and accumulator Stack←(A); Stack←(PSW)		
F5	PUSH PSW	None

Hex	Mnemonic	Flags
Pop register pair (rl)←((SP)); (rh)←((SP)+1); (SP)←(SP)−2		
C1	POP B	None
D1	POP D	None
E1	POP H	None
Pop PSW and accumulator (A)←Stack; (PSW)←Stack		
F1	POP PSW	None

Hex	Mnemonic	Flags
Exchange stack top with H and L (L)⟷((SP)); (H)⟷((SP)+1)		
E3	XTHL	None
Move HL to SP (SP)←(H)(L)		
F9	SPHL	None

Hex	Mnemonic	Flags
Input (A)←(data)		
DB	IN data	None
Output (data)←(A)		
D3	OUT data	None

Hex	Mnemonic	Flags
Halt		
76	HLT	None
No operation		
00	NOP	None

Hex	Mnemonic	Flags
Enable interrupts		
FB	EI	None
Disable interrupts		
F3	D1	None
Read interrupt mask		
20	RIM	None
Set interrupt masks		
30	SIM	None

Bibliography

ARTWICK, B. A. *Microcomputer interfacing*. Prentice-Hall, 1980
BARTLE, T. C. *Digital computer fundamentals*. McGraw-Hill, 1977
BOYCE, J. C. *Microprocessor and microcomputer basics*. Prentice-Hall, 1979
CHEN, Y. *Digital computer design fundamentals*. McGraw-Hill, 1962
CLAYTON, G. B. *Data converters*. MacMillan Press, 1982
COLIN, A. *Programming for microprocessors*. Newnes-Butterworth, 1979
DAVIES, D. W., BARBER, D. L. A., PRICE, W. L. and SOLOMONIDES, C. M. *Computer networks and their protocols*. Wiley, 1979
ECKHOUSE, R. H. and MORRIS, L. R. *Minicomputer systems*. Prentice-Hall, 1979
FREEDMAN, M. D. and EVANS, L. B. *Designing systems with microcomputers*. Prentice-Hall, 1983
GAULT, J. W. and PIMMEL, R. L. *Microcomputer-based digital systems*. McGraw-Hall, 1982
HILBURN, J. L. and JULICH, P. L. *Microcomputers/microprocessors*. Prentice-Hall, 1976
HILL, F. J. and PETERSEN, G. R. *Digital systems, hardware, organisation and design*. Wiley, 1978
HOLDSWORTH, B. *Digital logic design*. Butterworth, 1982
INTEL *MCS-80/85^{TM} family user's manual*. Intel Corporation, 1980
INTEL *8080/8085 assembly language programming manual*. Intel Corporation, 1977
KLINE, R. M. *Structured digital design*. Prentice-Hall, 1983
KLINGMAN, E. E. *Microprocessor system design Vol. 1*. Prentice-Hall, 1977
KLINGMAN, E. E. *Microprocessor system design Vol. 2*. Prentice-Hall, 1982
KORN, A. G. *Microprocessors and small digital computer systems for engineers and scientists*. McGraw-Hill, 1977
LEWIN, D. *Theory and design of digital computers*. Nelson, 1972
MANO, M. M. *Digital logic and computer design*. Prentice-Hall, 1979
MANO, M. M. *Computer system architecture*. Prentice-Hall, 1982
MILLAM, J. *Microelectronics*. McGraw-Hill, 1979
OBERMAN, R. M. M. *Digital circuits for binary arithmetic*. Macmillan, 1979
OGDEN, C. A. *Microcomputer design*. Prentice-Hall, 1978
PALMER, D. C. and MORRISS, B. D. *Computing science*. Arnold, 1980
PEATMAN, J. B. *Digital hardware design*. McGraw-Hill, 1980
SHARP, K. L. *Microprocessors and programmed logic*. Prentice-Hall, 1981
SHORT, K. L. *Microprocessors and programmed logic*. Prentice-Hall, 1981
TAUB, H. *Digital circuits and microprocessors*. McGraw-Hill, 1982
TOCCI, R. J. *Digital systems*. Prentice-Hall, 1980
TOWNSEND, R. *Digital computer structure and design*. Butterworth, 1982
WAKERLEY, J. F. *Microcomputer architecture and programming*. Wiley, 1981
WISTROWSKI, C. A. and HOUSE, C. H. *Microcomputer architecture and programming*. Wiley, 1981

Index

Access time, 66, 109
Addend, 7, 20
Addressing modes
    direct, 135
    immediate, 134
    implied, 133
    indexed, 137
    modified page zero, 139
    register, 135
    register indirect, 135
    relative, 138
Algorithm, 17
Analogue switch
    bipolar, 298
    MOSFET, 294
Analogue-to-digital conversion
    bipolar, 301
    characteristic, 309
    coding, 299
    errors, 322
    post-subtractive, 318
    pre-subtractive, 318
    quantization noise, 309
    quantum, 309
    successive approximation, 316
    transition points, 309
Analogue-to-digital converters
    counting, 314
    parallel, 320
    software, 323
    successive approximation, 318
    tracking, 316
Architecture, microprocessor
    accumulator, 94
    arithmetic/logic unit, 96

Architecture, microprocessor (*cont.*)
    general storage registers, 95
    index register, 95
    instruction register, 93
    program counter, 92
    stack, 94
    stack pointer, 94
    status register, 96
Arithmetic circuits
    controlled adder, 58
    4-bit adder, 56
    full adder, 55
    true/complement unit, 57
Arithmetic/logic unit, 59-61
Arithmetic overflow, 7
ASCII code, 247
Assembler
    directives, 194-197
    location counter, 198-199
    pseudo-instructions, 194
    two-pass, 192, 198-201
Assembly language fields, 190-191
Assembly manual, 192

Baud rate, 246
Base address, 137
BCD arithmetic
    conversion to binary, 23
    signed, 22
    10's complement, 21
    unsigned, 19
Bipolar technology, 69, 71, 76
Bit rate, maximum allowable, 254
Booth's algorithm, 16
Bootstrap program, 215

335